FAST N' LOUD™

BLOOD, SWEAT AND BEERS

RICHARD RAWLINGS
WITH MARK DAGOSTINO

WM
WILLIAM MORROW
An Imprint of HarperCollins*Publishers*

HarperCollins books may be purchased for educational, business, or sales promotional use. For information please e-mail the Special Markets Department at SPsales@harpercollins.com.

A hardcover edition of this book was published in 2015 by William Morrow, an imprint of HarperCollins Publishers.

FIRST WILLIAM MORROW PAPERBACK EDITION PUBLISHED 2016.

Designed by Jamie Lynn Kerner

Library of Congress Cataloging-in-Publication Data has been applied for.

ISBN 978-0-06-238787-5

16 17 18 19 20   OV/RRD   10 9 8 7 6 5 4 3 2 1

Any success in life comes down to three things, and as such, I dedicate this book to:

1. God—for giving us America, the greatest country in the world that allows me to do what I do!
2. Family—namely my wife, Sue, who stood by me all these years, in success and in failure. And . . .
3. Friends—especially my best friend, Dennis Collins, who took me under his wing and taught me the car business, and who has always been there with a cold beer.

CONTENTS

COURTESY OF DISCOVERY COMMUNICATIONS.

INTRODUCTION

Whooo!

Want to know what it feels like to do 150 MPH on a long stretch of highway? I'll tell you this: it's nowhere near as cool as going 207 MPH on that same stretch of road.

There are a lot of people out there who've topped 100, and quite a few everyday guys and gals who've topped 120 or 140 in a street-legal vehicle on a public highway. But the difference between driving 150 MPH and driving 200 MPH is as big as the difference between doing zero—like, actually standing still—and doing 100. The roar of the engine and wind and road comes pounding through your body as the trees and lights and lampposts and even other vehicles become blurs of color and the adrenaline kicks every one of your senses into the highest gear possible.

It's f—kin' awesome!

Speed that close to the ground is scary as hell, though, so "Don't try this at home, kids!" Going that fast without killing yourself or anybody else requires an unbelievable amount of attention, intensity, and focus. But if that's what the job calls for, that's what I'm gonna deliver. And topping 200 MPH in order to break a world record was exactly what the job called for on one kick-ass trip back in 2007 when my buddy Dennis Collins and I answered a bet and teamed up for the rally of a lifetime: the Cannonball Run. It was right in the middle of that run, on a lonesome stretch of Interstate 40, where the speedometer hit 207 MPH in Dennis's 1999 Ferrari 550 Maranello—and we kept it over the 200 mark for a whole lotta miles in order to make up for some lost time along the way.

Just about everybody's heard of *The Cannonball Run*, the 1981 movie starring Burt Reynolds, Farrah Fawcett, Dom DeLuise, and a bunch of other big stars from that era. Well, that film was based on a real road rally called the Cannonball Run that was first run back in the 1970s, following a cross-country route from Midtown Manhattan to the Portofino Inn and Yacht Club in Redondo Beach, California. There's been a tradition of less-than-legal road rallies all over the U.S. and Canada ever since, and I've run in plenty of them, but the Cannonball is the most legendary. Dennis and I set out to beat a cross-country record that had stood for more than twenty-five years: The original Cannonball winners made the Run in thirty-two hours and four minutes. No one had been able to touch that record in all that time, and a lot of people thought it was impossible to beat. In fact, a buddy of ours, Jay Riecke, bet us $50,000 that we couldn't do it, and that was all it took for Dennis and me to give it a go.

Armed with a stack of cash and a supply of beef jerky, we took off from the Red Ball Garage at 8:30 P.M. and rocketed toward New Jersey. We knew Manhattan traffic was one of our biggest obstacles, so I called ahead to a buddy of mine named Nassar who owns a limo company. We hired his whole fleet of drivers to get in front of us and briefly block traffic onto the bridge so we could get through without a hitch. We had drivers run ahead of us through New Jersey, too, driving fast and weaving around so that any cops in the area would pull them over and be fully occupied when we came speeding past them.

For those who might be reading this with horror, thinking how terrible and dangerous it is to drive at high speeds on public roads, I just want to acknowledge that I didn't take it lightly. A lot of guys had done this before us, and we were very safe about it. We thought about where we were, we didn't goof around, and we weren't running people off the road or acting like idiots. We were very respectful about what we were doing at every turn. Realize that a lot of people drive 75 or 80 on the highway as it is. Our average speed over the course of this whole route was 87.6 MPH. Not exactly crazy driving.

We dropped down along the original route of the Cannonball all the way to Interstate 40 and then beat it west just as fast as we possibly could, only stopping five times for gas. Good thing we limited our meals to beef jerky and energy drinks, because those were the only bathroom breaks we made the entire time, too!

I was married then, and my wife, Sue—with less than a day's notice—flew out to Redondo Beach and grabbed a hotel room at our destination. She's a computer whiz, and she actually created a computer model for us that would figure out how fast we needed to go in order to break the record. By checking in via cell phone with our location and time, she could tell us what our estimated finish time was, and how fast or slow we needed to go in order to make up for lost time. She also told us when we could hang back a bit so we wouldn't get stopped. After all, the time it would take to suffer a single speeding ticket along the way could've blown the whole adventure.

Dennis and I went a little nuts on the cop watch. This whole endeavor wasn't exactly legal to begin with, right? So we pulled out all the stops. We had radar detectors and diffusers and jammers, everything in the world. We even had an Opticom, which is the device that changes stoplights from red to green when you're on your way to the hospital in an ambulance and what have you. Believe it or not, it's not illegal to own one of those devices. You can buy 'em on eBay. It *is* illegal to *use* one, so I'm not gonna say here whether we used it or not!

The first half of the trip was exhilarating but exhausting. I think the only thing that got us through was the adrenaline and the camaraderie of making that crazy drive with a good friend. Once we got about halfway through, though, we started to feel good. We got lucky. We had good weather. It was Mother's Day weekend, so we weren't hitting a lot of traffic. I mean, we really thought we might pull this off!

Fast N' Loud wasn't on TV yet, but Gas Monkey Garage had been up and running for a few years by then. Aaron Kaufman and I had a pickup and trailer that we drove to rallies all over the country, trying to let people know what we were all about. So when this thing started to look like it might happen, I called Aaron down at the garage and told him to drop everything he was doing.

"Drive to Redondo Beach right now, man! Go!"

I figured Dennis and I might make a splash in the press if we could pull this thing off, and having our Gas Monkey logo on the side of a truck parked right behind the Ferrari at the finish line certainly seemed like it would be good marketing.

I could hardly believe it, but as we reached California, it looked like we were actually going to do it. We were set to beat the thirty-two-hour mark! There was just one problem: we miscalculated the Ferrari's fuel needs. We must've burned off a ton of gas during one of our 200-MPH stretches, because just as we hit the outskirts of L.A., we realized we were running on fumes.

Pulling off the freeway and stopping at a gas station for even a few minutes might have done us in. We knew that. We flipped out! *How could we get so close and then blow it with only twenty miles to go?*

That's when I remembered: Aaron was on his way to Redondo Beach.

I hopped on the cell phone. "Aaron, where are you?"

Lo and behold, he wasn't all that far ahead of us.

"Get off the highway and get a five-gallon bucket and fill it up with gas," I told him. With my wife's help we picked a mile marker and had Aaron sit on the side of that road with that bucket of gas until we got there.

Finally we saw Aaron up ahead. *"Whoo-hoo!"* I yelled as we slid to a stop and he dumped that gas in like one of NASCAR's finest pitmen. *Boom!* We were off and running again on the bad-ass adrenaline high of knowing we were about to make history.

Weaving through traffic and carefully blowing through stop signs, pushing right to the end—we did it. Sue was waiting at the curb watching the official time as we roared into the hotel parking lot at thirty-one hours and fifty-nine minutes flat. A new world record for the Cannonball Run!

Aaron pulled up in the truck a few minutes later, and, just as I suspected, our feat got picked up in all of the automotive press. I was so thrilled, I shouted it from the mountaintops, man. I even caught the attention of a certain car buff named Jay Leno, who called us up and had us come on his show to share our accomplishment with the world. Remember, *Fast N' Loud* wasn't on TV in 2007. I was just some guy with a garage down in Dallas! Getting on a major TV show and getting press coverage like that was huge for my growing business!

When Dennis called me up in 2012 to remind me it was the five-year anniversary of our Cannonball adventure, I walked right into a tattoo parlor and had them commemorate that feat permanently on my left forearm. It's right there: "31:59," inked into an image of a stopwatch forever on my skin.

Dennis and I broke the longest-standing Cannonball Run record, and there is no one on earth who can take that away from us.

Some people live by a "Go big or go home" philosophy. But for me, "Go big or go home" was never enough. Go biggest, go baddest, go raddest, go for broke, go for everything. That's the way I live. That's what brought me to TV. That's what built the growing Gas Monkey empire, from Gas Monkey Garage to Gas Monkey Bar N' Grill to Gas Monkey Racing and Gas Monkey Apparel . . . heck, by the time this book comes out, you'll be able to buy Gas Monkey–branded tequila at bars all over the Western Hemisphere!

If you've picked up this book, chances are that you or someone you know is a fan of my show on Discovery Channel called *Fast N' Loud*. We've been on the air since 2012, but I've been flipping cars, building hot rods, and tearing up the streets for a whole lot longer than that—and I'm about to show you what I mean. The idea here is to let you run backstage when the security guard's not looking to get a glimpse at what I'm up to behind the scenes. While we roll, I'll share the backstory of where I came from and how I got here, and I'll give you a little glimpse of where I'm going, too.

I'm going to talk to you about the cars I've loved, the women I've loved, the traveling I've done, and the crazy twists and turns along the way as I built the Gas Monkey brand and reputation to the point that made *Fast N' Loud* explode in popularity.

Then I'm gonna take you through some of your favorite builds as I reminisce about some of the most popular episodes of the show you love, as well as my personal favorites. And finally, I'm gonna share a few tricks of the trade when it comes to flipping cars. Here at Gas Monkey Garage, we pull in a whole lot of cash every month just by buying and selling some really cool vehicles without hardly laying a hand on 'em. The thing I've found when it comes to cars is that if you love it, if you're dedicated, if you're smart and fast and make some noise, there's no reason you can't go out there and flip a few of 'em yourself. I mean, who wouldn't want to make themselves some extra income while having a good time? Isn't that what we all want in life? Cars are fun! Money's fun! GYSOT, man: *Get you some of that!*

Hell, if you really want to get dedicated, you can go out and make one heck of a living flipping cars. I know it's possible because I've done it. Not only that, but I managed to turn my passion for hot rods into one of the biggest, most successful car shows on all of television—a feat that everybody told me couldn't be done.

So hop on in, buddy. I'm about to drop the pedal down and burn some rubber in the parking lot before we head out onto the open road, and I couldn't be happier to take you along for the ride. *Whooo!*

PART ONE
THE "AUTO" BIOGRAPHY

Me in my '68 *Thomas Crown Affair* Shelby. *COURTESY OF DISCOVERY COMMUNICATIONS.*

Me as a youngin'. *COURTESY OF RICHARD RAWLINGS.*

FIRST GEAR

I was born a poor black child . . ."

Ha! I always wished I could start my autobiography that way. That has got to be one of the funniest openings from any movie in history. I suppose you want to hear the story of my own upbringing instead of a fictional one written by Steve Martin, but if you watch the show, you already know that I'm a fan of all sorts of great flicks from the seventies and eighties. Steve Martin's *The Jerk* holds a place right up there with the likes of *Ferris Bueller's Day Off*, *American Graffiti*, and of course *Smokey and the Bandit*, the king of the revved-up car-centric films that rolled into theaters back when I was a kid. I'll never forget what it felt like to sit in a movie theater and watch that black Trans Am tear up the streets. All I kept thinking was, I gotta get me some of *that*!

Well, the story of my childhood might not be quite as riveting as some fast-moving Hollywood film, but I think it will give you a peek at where the earliest seeds of the Gas Monkey empire were first planted. So here goes.

I was born in Fort Worth, Texas, in the infamous all-American year of 1969. Blame it on the free-love era, or something like that, but I was basically the product of too young a mother and too young a dad—a situation that led to a bit of a rough patch in my early years. My mom left us when I was two years old.

My dad got stuck quitting high school in order to raise me and my older sister, Daphne. (You recognize Daphne from the

That's me as a baby, with my big sister, Daphne. *COURTESY OF RICHARD RAWLINGS.*

show. She's my chief accountant and all-around ballbuster these days, and has been pretty much from the start.) The thing is, my dad should have been out there living it up and having a good time like other young dudes in the 1970s. The first apartment we all lived in was like some swinging bachelor pad, complete with shag carpet and fishing nets hanging from the ceiling. A bunch of liquor bottles filled with colored water on the windowsill were the fanciest decorations we had. He quickly found that raising two kids put a major damper on his bachelorhood, though, and he was forced to straighten up his act real quick. The truly amazing and inspirational thing to me is that he did. He grew up fast. He put us first. He started working to support the three of us, and when I say work, I mean *work*.

There's not a time I can remember before I was old enough to leave the house when my dad wasn't holding down two or even three jobs at once to pay the bills and put a decent roof over our heads. His primary job was down at the big local grocery store, where he landed a job as the produce manager. My sister and I would go down there with him on Saturdays and hang out at the store all day long because there was nowhere else for us to go. The job I remember most, though, was the one he took delivering newspapers. I remember it because I had to work that job with him.

From the time I was about seven until I was probably seventeen, my dad would wake me up at three or four in the morning every day. He would put me in the backseat of his car surrounded by stacks and stacks of newspapers that needed delivering, and my job would be to roll 'em all up, one by one. I would roll the newspapers for him and put them in the front seat, and he would throw them out the windows to all the houses.

We'd get done with that at about five or six in the morning

and I'd be able to lie down for thirty minutes to an hour before I had to get up and go to school while Dad went off to spend his day tending to lettuce, tomatoes, and, I suspect, the occasional lady shopper. I didn't have one of those PTA-type moms dropping me off at the front door of school every day, either. My dad remarried, so we had a mother figure at home. She just wasn't the type of woman who'd baby us. Daphne and I both had to walk to the bus stop no matter what the weather was and then ride on one of those rickety yellow busses full of screaming kids. Then I had to race home from the bus stop at 3:30 when I got out of school, because back then there was a morning paper and an evening paper. So my dad and I had a paper route to get to in the afternoon, too. I wasn't able to go play with my friends or start my homework or anything until about six o'clock, when our work was finished.

I suppose there's no question where I get my work ethic: it's from watching my dad. But I think that really led to another driving force in my life, too: I knew that I never wanted to have to work to the bone like he did just to make ends meet. I'm not afraid of hard work. Don't get me wrong, I really, really respected my dad for doing it, but I didn't want to spend my whole life working just to pay the bills and never really get ahead. My dad didn't want that life for me, either. The three strongest principles he demanded of me were to be respectful of my elders, to be respectful of my family, and to work hard. Most of all, I think he instilled in me that he didn't *want* me to have to work that hard. He thought school was important, and he thought that getting a good job with good benefits was important. I think he felt that he wasn't able to do that because he didn't graduate high school, so he insisted that Daphne and I stay in school and strive for something better.

I wouldn't say we were on the super-poor side, but despite all of his hard work we grew up kind of below average. We rarely had anything "extra." He couldn't go out and buy us the latest toys and electronics when they first came out. We couldn't go out to nice restaurants or buy the nicest clothes. We were lucky to have what we had, and the biggest thing we had is what I would consider a normal house with three bedrooms and two bathrooms in a lower-to-middle-class Fort Worth neighborhood. My dad took real pride in absolutely every part of it, too. He mowed the lawn perfectly, and even edged the grass all around the yard. We kept a clean house. We even helped him keep the garage immaculate. I suppose that rubbed off on me, too. To this day I take great pride in everything I'm fortunate and lucky enough to own because I saw him take pride in, care for, and maintain whatever it was he was able to purchase and keep for his family back then.

That included his cars. My dad had a hankering for cars from the start, and I think that was really the only area where he would treat himself to a little something special. Some of my earliest memories are from right out in the driveway with a bucket and sponge, helping him wash the fenders to keep his car all shiny.

My dad didn't really have a lot of money to spend on cars or motorcycles, but he always had something. I mean, it wasn't expensive. It wasn't the best one out there. But it was his. It was his toy.

Dad wasn't a tinkerer. He didn't know how to get in there and fix the engines or turn a junker into a hot rod or anything. He could change the oil, and he knew what he liked as far as choosing new wheels and making it look really good. But he always had a fun car that was sort of a showpiece, something we could take out for Sunday drives, and he took care of 'em like they were another child or something.

He was so proud of those cars that he tried to get them into car shows forever. But they were never quite up to snuff. That is, until he purchased a '65 Mustang 2+2, maroon with black interior.

Someone else must've dropped out at the last minute or something, because the night before the big Autorama show at Market Hall in Dallas, they called him up and said, "Hey, you're in. But you've got to load in first thing in the morning."

We didn't have a truck or a trailer to haul that Mustang in, so we had to drive the car to the show in the pouring rain. As soon as we were in the arena my job was to take a one-gallon bucket and run back and forth from the bathroom so he could rinse off a rag and keep wiping the car down over and over until the water spots were off of it and it looked just right.

It's funny looking back on it, because I don't remember a lot of the specifics about the other cars he had. I don't remember much about the specific motorcycles, either, yet I know he had a few. Maybe my lack of memory is because he never let me touch them on my own. Hell, he would yell at me if I rode my bike between his cars in the driveway, afraid that I'd scratch one of 'em with the handlebars or something. I know now what he was really worried about: he knew that he didn't have the money to get it fixed if something like that happened.

I do remember he had a custom truck back in the seventies, with the shag carpet and a bed in the back, and he would take his shoes off and make us take our shoes off before any of us got in it. He was passionate and strict about all of his vehicles. Hell, he was passionate and strict about lots of things, including discipline. He was quick to the belt when we got out of line, and as you can probably guess from my personality, I was the type who got into all kinds of trouble every chance I got. So I got to know his belt pretty well.

My dad was a kick, too, though. He'd do some crazy things. Like one time when the school was throwing a fund-raising fair, he rode his motorcycle up there with his leather jacket on and opened up a kissing booth. Everybody else baked cookies and sold their kids' hand-drawn art and clay ashtrays, and there's my dad, like, "Where's all the moms at? One dollar per kiss!" People in Fort Worth still talk about it today!

Other kids might've been embarrassed by my dad's over-the-top style of doing things, but not me. I dug it. I thought he was the coolest. And you know what? He was. He set the bar high—and from the moment I hit my teenage years, I was determined to not only reach that bar but to jump right over it. It sure would take me a long time, though. I was skinny and pretty geeky and wasn't too good with the ladies back in high school, if you can believe it. Looking back on it, I'm glad I didn't waste my cool in high school, though. When you go to just about any high school reunion a lot of the guys that were like me—a little too thin, a little bit of a late bloomer, a little shy with the girls—they're pretty badass now. They've got the good job and the hot wife and all that. And some of the jocks that were getting all the action and thought they were the s—t in high school are kind of fat and bald and not doing so well. As they say, karma is a bitch!

High school was actually pretty rough on me. I didn't fit in well, and I certainly didn't like it. I think mostly it was just too slow for me. The thing I tuned in to early on, though, was that the easiest way to be cool was to stack up a whole lotta cash.

I loved money. I always did. I still do! But it was in my early teenage years when I really got a hankering for wanting to *make* money. My *own* money. And it was outside of school where I seemed to get my best education.

I worked all the typical teenage jobs. I worked at the burger joint. I worked at the drugstore. I worked everywhere. But it wasn't those jobs that taught me about making money. It was wheeling and dealing cars on the side as a hobby.

A lot of kids get a car in high school. Sometimes you'll find a kid who sells his car and buys another one, or who buys some hunk of junk and fixes it up in his dad's garage and starts showing off doing donuts with his girlfriend in some parking lot. I wasn't that guy. I was something a little different.

I would estimate that I owned a series of nearly twenty cars just in high school.

From the moment I got behind the wheel, I made it my goal to trade up to something better, or to make a few changes that could make me a little money so I could go and buy a different one, just for fun.

I had a few motorcycles mixed into that twenty, too, and to me, the wheeling and dealing was almost as much fun as driving 'em. Maybe more!

In the state of Texas back then, as soon as you turned fourteen, you could get something called a hardship license. Because my dad and my stepmom both worked, I was able to get my license in order to drive to my job. That was the "hardship." *Boom!* I couldn't afford a car at that age, so I got a motorcycle license instead. Suddenly I was off and running at the age of fourteen.

My very first car was a 1976 Chevrolet Impala. My dad bought it for me shortly before I turned sixteen. Much to his surprise, I found another car I liked better not long after that Impala landed in our driveway. I went and sold it for a hefty profit and bought myself a '74 Mercury Comet. It was sort of a piss-color green, with a green interior. For some reason, I've always liked green. I was hooked on it! I put the stripes on it, and new wheels, and fresh tires, and took some of the money I was making and put a big stereo into it and all that stuff. Then I suddenly realized it was worth quite a bit more than I'd paid for it.

Boom! I sold it, bought another car, and stashed a little chunk of cash away in the bank to use for the next one.

When I started all that wheeling and dealing, I found I had a knack for keeping track of money and figuring out what was profitable and what wasn't. It was strange in a way, because in school I had always done terrible in math. Something changed in high school, though, thanks to my shop teacher, of all people.

These days, occasionally somebody will ask me, "How can you be successful in business if you weren't very good in math?" And I say, "Well, I wasn't very good in math at first. But then my shop teacher at Eastern Hills High School in Fort Worth would come in every day and he'd write a number on the board. And it was usually like $4.30, or $6.10, or whatever. And he'd be like, 'You see that? That's what I made last night while I was sleeping, because I got a retirement account and I put money away. And that's what you kids need to be focusing on. You get out there and you take a portion of everything you make, and you put it up.' He drilled that into our heads."

Besides that important lesson in saving, I wound up talking to him one day about how terrible I was at math. I said, "I just can't get it."

He goes, "You like money?"

"Yes, I love money," I said.

He goes, "Well, every single math problem from this point on, whatever they tell you—eight plus four times ten, or whatever—just put a dollar sign in front of it."

I said, "What do you mean?"

He said, "It's just numbers. If you like money, and you can count, when they give you a math problem, put a dollar sign in front of it."

I swear it was like a big fat lightbulb turned on over my head, just like in one of those old cartoons. From that point forward, I never had a problem. In fact, I'm so good at keeping track of the math in my business, it drives my sister nuts. Daphne has run the accounting at pretty much all of my companies through the years, and there will be times when she'll show me the books and I'll turn to her and say, "That's not right. There should be like fifteen hundred more in that account." She'll go away and look at it and realize that something was placed in the wrong column or something. She hates admitting that I'm right, but it happens all the time! I do this with accounts I haven't seen in months. I just keep track of it all in my head. It's just numbers—and as long as there's a dollar sign in front of it, I can track it.

By the time I got ready to graduate high school (which I barely did), I was already making pretty good money buying and selling cars on the side. In fact, as graduation neared, I'd already traded my way into owning one of my dream cars: a '78 Trans Am. It was red, but otherwise it was basically the Bandit Car from *Smokey and the Bandit*. Black interior, a big eagle on the hood, T-top, four-speed . . . It was just *bad*.

I paid $4,500 for that car. That was a lot of money back then. But it was the right car for me in that moment. When I told my dad I'd found one and was fixing on buying it, he said I couldn't have it because we couldn't afford the insurance. I completely ignored him. I came home with it and revved that fat V-8 in the driveway. Man, was he pissed off.

"What the hell did you do?" he yelled.

"Well," I said, "I paid cash for the car out of my own pocket, and I already bought a year's worth of insurance. So what's the problem?"

"Well," he said, just shaking his head. "There's not one."

So there I was, all of eighteen years old, kissing high school and all of its stupid problems good-bye and driving off into the sunset in a '78 Trans Am. I was riding high.

That's when I witnessed something awful.

I was getting ready to move out of the house when the grocery-store chain where my dad worked went belly-up. Just like that, he was out of a job. The economy was a mess then, too, so that

Dad and me . . . *COURTESY OF RICHARD RAWLINGS.*

chain's retirement program failed on top of it all. After fourteen, maybe fifteen years on the job, my dad was out of work and lost all of his retirement savings in one fell swoop.

He would eventually recover. He was always a hard worker. But things got pretty scary there for a while. Seeing something like that happen to my dad made me think hard about the type of job I wanted to have in life, and how important it would be to find a career that wouldn't just up and disappear one day. It also taught me pretty quickly that I'd better make my own way in life, and I'd better not rely on somebody else's potentially faulty business model to provide my security and my future.

REVVED UP

Guess what company I landed a job with right out of high school?
Miller Lite.

No joke! In case you haven't noticed, I've been a spokesman
for Miller Lite since shortly after *Fast N' Loud* hit the airwaves.
It's the only beer I drink. I keep a refrigerator full of it in the back
corner of Gas Monkey Garage at all times. People see my face on
Miller Lite billboards and cruising down the road on the side of
Miller Lite trucks all over the place, but back then, my first job
out of high school was driving a Miller Lite keg truck at night.
Bars tended to be smaller in the 1980s, which meant that they
didn't have a lot of room to store extra kegs in the back. So I basically drove an emergency keg-truck route: when a bar would run
out of Miller Lite, they'd call me. My pager would start beeping
and I'd haul my butt over to that bar to get them a new keg.

One night I wound up delivering to a bar in West Fort Worth.
It was late at night, and one of the locals who was often there, an
older gentleman, started giving me a hard time.

"What are you going to do with your life?" he said. "You know you can't drive a keg truck forever."

I thought, *You're sittin' in a bar, so what the hell is your problem?* But what I said was, "Well, sir, my dad always told me I needed to get a good job with good benefits. So quite frankly, I've been thinking I might go be a police officer."

"Well, then why aren't you in the police academy instead of delivering beer?" he asked.

"It's not that easy. You've got to be sponsored or hired by the city before they'll let you enroll," I told him. It was true. I'd already started looking into it. Cops make good money, drive fast cars, carry a firearm, and have just about the most secure retirement funds around. Seemed like a good deal to me.

"Well, why didn't you say so?" he said. "*I'll* sponsor you!"

I thought this guy was blowing steam, but guess what? He was the mayor of a nearby city! And he kept his bar-stool promise. Three weeks later, I was enrolled at the police academy.

Talk about a life lesson! I could've blown that guy off or mouthed off to him, but I didn't. I always showed respect to my elders, just like my dad always taught me, and late one night in a random bar it paid off for me. Big-time.

I became an officer before I was twenty years old. My dad was really proud. He could hardly believe it! I had already moved out and had my own place, so I kept driving the keg truck as a way to pay my bills the whole time I was in training. Sure enough, though, I became a cop in the city of Alvarado. I didn't stay too long. When the opportunity became available I jumped over to a slightly higher-paying job as a Tarrant County constable (which is just another form of police officer), covering the whole Fort Worth area. The police work was only part-time, though. I was making enough money to keep up my penchant for buying cool cars and motorcycles on the side, but not much else. I wanted to find full-time employment.

I thought the answer would be stepping up to a full-time cop job with a big city, and the force that everybody wanted to work for at that time was in a suburb of Dallas called Coppell. Coppell was the fastest-growing city in the area and they had four or five openings that I knew of, so I picked up an application packet and filled it out. It was a huge packet full of background-check information and everything you can think of. It took me a while to complete, but when I dropped it off, I noticed another job posting on their bulletin board. It said they were accepting applications for new firemen, too. I'd never considered becoming a fireman before, but the application packet was exactly the same as the one I'd just filled out. So I Xeroxed it and turned them both in.

Next thing I knew, I was called in for some physical tests of strength and endurance. Luckily I'd grown out of my scrawny phase. I passed with ease and I wound up landing a job as a fireman, one of four full-time, salaried spots out of probably two thousand applicants. Life is crazy, isn't it? You just never know what'll happen if you put yourself out there and go for it.

So I'm a fireman and a part-time cop on the side. (Since firefighters work an odd schedule of one day on, two days off, I had time on my hands.) Oh yeah, and the firefighter training they put me through included full EMT training, too, so I picked up even more work as an EMT on the side.

There I was, twenty-one years old, and I was a certified police officer, firefighter, and EMT. No more driving a keg truck for me. And talk about cool. Remember how I said I wasn't really good with the ladies in high school? Yeah. Try putting on a uniform and see how fast your love life changes.

Having all of that success so quickly didn't settle me down, though. In fact, it fired me up. I got antsy. I wanted to see what else I could accomplish. My head started filling up with all sorts of crazy entrepreneurial ideas, and as far as I could tell, there was nothing stopping me from pursuing any of them. So that's what I did.

I opened up a detail shop with a buddy of mine. He knew all the ins and outs of the detail business at the time, and I knew what I liked after outfitting so many of my own cars through my high school years, so we blew it out and quickly built a following. I'd gotten lots of practice making cars look good, so I figured that was an easy way to make some additional cash on the side. The thing I found is that everything I did made me want to do even more. I developed a sensational appetite for success, and somewhere deep down, I knew that I wanted to build and do something big. *Really big.*

I would sit around the firehouse with the other guys and bounce things off the wall all the time. Crazy ideas for businesses and products, you know? My sister, Daphne, says I used to bring up all sorts of crazy ideas when I was younger, too, just sitting around the dinner table. I suppose I'd always had a bit of that imaginative, entrepreneurial drive in me, but this firefighting era of mine was the first brainstorming period that felt like more than a dream to me. I was positive these ideas I was tossing around were actually going to lead to something.

The older firefighters, who were all set in their ways and just waiting to retire, would laugh at me like I was some sort of dreamer and fool. I never understood that. Doesn't everybody dream of something bigger? Something more?

Finally, one day, I came up with an idea that would prove to all the doubters that I could accomplish something nobody'd ever thought of before.

I sat at that station and I told them, "When I got done having my car washed at the drive-thru this morning, they asked me if I wanted a litter bag. A litter bag! There aren't even any knobs in my car that a litter bag can hang from anymore. It's out of date, you know? It's like some sort of promotional tool from a bygone era."

It really was. Cars were changing. Handing out a little plastic litter bag that would hang from the headlight knob or the cigarette lighter just didn't cut it. It was useless to me. That got my mind turning.

"What I could really use is one of your towels," I told the gal at the car wash.

"Well, those are really expensive," she said. "We can't just give those away." I asked her how much they paid for 'em, and she told me they were a dollar and a half apiece. These tiny little terry-cloth or shammy towels or whatever they were.

As I was driving into work, I thought about it: the litter bag has been around since the fifties, and car washes use it to advertise their car-wash name, and the businesses around them pay to have their ads on those bags, too, which must be how the car washes offset their cost. They go to the businesses around them and say, "Hey, I'll put your restaurant name on there, Luigi," and all that kind of stuff. Then they give the bag away.

That's when it occurred to me: Why can't you do the same thing with a nice disposable printed towel?

The other firemen were like, "You're an idiot. That will never sell. You're not gonna print on a paper towel and make money doing that! Ha ha!"

Their skepticism spurred me on like I was a bull at the damned rodeo. I started asking around. I made a bunch of phone calls. I looked into whether it was possible to get a thick disposable towel made up, and how much printing area it would have, and how much they would cost. I really thought it might work. Then I found out there was an international car-wash association that had a giant show every year in Vegas, and I thought, *What the hell. I'm gonna try this.*

I printed up some sample towels for very little money, bought a small spot at that trade show, and drove all the way out to Vegas to see if I could sell a few. I'd never been to a trade show before and had no idea what I was in for. All these big companies were there, and everyone had spent a ton of money on all sorts of fancy displays. All I had was a table and a sign that said PROMO WIPES!

One of my original Promo Wipes car-wash towels. *COURTESY OF RICHARD RAWLINGS.*

I knew I needed to do something to draw more attention. It was a real turning point for me. I knew that if I didn't blow it out and do something over-the-top, I was gonna fail—and there was no way I was going back to the firehouse with my tail between my legs. So I called up one of the local modeling agencies the night before the show opened. I hired a couple of pretty girls to help show off the towels. I also went out and rented a Hummer H1, which was a pretty badass thing to do in the 1990s, and had it logoed up real quick. The morning the trade show started, I drove that Hummer up and down the Strip throwing out business cards, and then I made sure the girls got busy attracting all sorts of attention inside.

I hoped we'd take in forty or fifty orders just to get things started. I would've been really proud of that.

By the end of the convention, we'd taken in 980 orders. Nine hundred and eighty! From all over the country.

Suddenly, I had a new business on my hands. A booming business! I also had enough orders that I had some sway. I flew to Green Bay, Wisconsin, and cut a deal with Kimberly-Clark, the paper company. Basically, instead of paying them a certain price per towel, I made them an offer to buy time on their printing machines whether I bought a certain number of towels or not. I knew I had enough orders to use up all that time, so the result was it brought my cost per towel down significantly, which increased my profit in a big way. *Boom!* Remember now, I didn't have a college education. I just ran the math, did some numbers, thought about what I wanted and needed, and then made it happen. I can't imagine what that giant corporation thought of this kid coming in to cut a deal with them, but I wasn't afraid, and I wasn't leaving there without making a deal. Sometimes you've just gotta grab the bull by the balls, you know?

From there, I built my way up. I probably risked about $5,000 getting that company off the ground, and $5,000 was just about all the savings I had at that time. I believed in myself and my idea enough to take that risk, and it paid off. I grew that side business into a nice little company, and on top of my firefighting and police and EMT work, I was making a nice living. I started riding a Harley during that time period. I went out and bought a gold-colored 1965 Mustang—a full-on Steve McQueen–style 2+2 fastback. One of the coolest cars I've ever owned. Life was good, man!

Funny thing about life, though. If you're not careful, sometimes when you're riding that wave, it can crash on you.

GUNS & MONEY

I've always had a phobia of driving in the rain. Especially at night, and especially if the night involves any kind of partying or drinking. That's just stupid, and I try my best to be smart about things. I try not to put myself in harm's way. I try to think things through as a general rule, but whether it's rational or not, I think driving in the rain is just asking for an accident.

The crazy thing about that phobia is that it may well have saved my life one time, in a way I never could have imagined.

So, it was pouring rain when my closest buddies, Ted and Scott, invited me to go party with them at a big club in Dallas. "No way, guys," I told 'em. "Have fun." It didn't make sense to me that we'd go all the way to Dallas just to grab a few beers and try to meet some girls. We'd have to drive all the way back, and that just seemed like a waste of time. The rain just made it a 100 percent no-go.

So they went. I stayed home. Later that night, I got a phone call that shook me up really bad. The story went like this: apparently the club wasn't all that happening that night. So Ted and Scott walked in, had one beer each, turned around, and headed right back to the parking lot. That's when they saw someone breaking into their car. They'd driven the show car from Scott's stereo business, which had the shop's name on the back window and everything. Clearly the thief thought he was in for a big score with all of that expensive stereo equipment in there.

"Hey! What the f—k do you think you're doing?" they yelled as they ran over to beat the crap out of the guy. That's when the thief popped up with a gun in his hand. He pointed it right at them. Scott ducked. The gun went off. Ted got shot in the face. He died right there. One of my closest friends. Just like that. Gone.

It's hard not to think about what might have happened if I'd gone with them. I made it a rule to never carry my gun when I was out partying. Having a firearm on you when people are drinking is just a dangerous thing to do. So it damn well could've been me that got shot. It shook me up to think about it, but I was nowhere near as shaken as Scott. He was in really bad shape as we buried our friend. I can't even imagine how it felt to be right there when it happened. There's such a thing as survivor's guilt, too—that nagging, terrible question about why he made it but Ted didn't.

I hated seeing Scott so torn up. Finally, maybe two or three weeks after the funeral, I talked him into going out: "Hey man, you gotta get out of the house." I'd talked to him about life's twists and turns, how s—t happens, how it wasn't his fault. But what he really needed was to get out of the house, have a few beers, find himself a piece of ass, and get back to living.

So we went out partying and Scott had a few drinks and loosened up. By the end of the night I was pretty loaded, and Scott was in no shape to drive, so I took a cab home from the bar. It was 1:30, maybe two o'clock in the morning by the time I got back to my house, and by then I'd decided I was hungry. It wasn't too smart of me, but I decided to jump in my prized gold '65 Mustang fastback and head to a burger joint down the road. It was only three blocks away. I didn't expect any trouble with the drive, and I really, really wanted a burger and some fries. (Sometimes I wonder if I was singlehandedly trying to define the phrase "young and stupid.")

I pulled around into the drive-thru lane and I noticed that the girl at the window was talking to three shady-looking characters. I was off duty and I'd had a few beers, but I still had my cop eyes on. I noticed those guys look back at me before they moved away from the window. I rolled up there, gave the girl my order, and handed her a twenty-dollar bill. She closed the window. I heard a click. She locked it. Right then and there I knew something was going down.

I looked in my rearview mirror and saw those guys coming up behind me on foot.

Instinctively I went straight down to where I usually keep my gun beside the seat. My left hand was still on the steering wheel. My window was open. One of the guys reached in to try to grab the keys out of the ignition as I realized my gun wasn't there. From that bent-down position with no visibility at all I slammed it into first and took off. That's when I heard the gunshot. I felt my car go crashing over the hedges and the curb. I heard more gunshots: *Pop! Pop! Pop!* I sat up in the seat so I could see. I tried to turn the wheel as I jammed it into second, and I couldn't understand why it wasn't turning. Then I looked over and saw that my left arm was just hanging.

I'd been shot.

"F—k!" I shouted.

I let go of the shifter and drove with my right arm. Then my left arm kind of came back to life and this burning pain seared in a trail from my shoulder down through my triceps to my elbow.

Tires screeching, I yanked the car into my driveway and ran inside, scaring the hell out of my sleeping roommate.

"What the f—k?" he yelled as he caught sight of the blood.

"Motherf—ker! I've been shot!" I yelled as I made a mad dash for my guns.

He picked up the phone and dialed 911, but I'd already decided that I wasn't gonna wait around while those guys got away.

"Man, f—k them. Let's go."

I wrapped my shoulder and arm in a T-shirt and I jumped back into my Mustang to go out looking for them, which probably wasn't the smartest idea. My roommate jumped in with me. It had always been a lousy neighborhood for as long as I could remember. It was so bad, I used to ride my Harley up over the curb going into my yard, kick the front door in, and park in the living room rather than park my bike outside. There was just too much of a chance that somebody'd steal it otherwise.

Our search was fruitless. By the time we got back to the house the police were there with the whole street blocked off.

We told them who we were, and the guys from the ambulance started to take care of me. That's when a cop I knew said, "Man! What happened?".

"Haven't you gotten a call from the burger joint yet?" I asked.

"What are you talking about?" he said.

"These dudes blew up with two guns and shot my ass and I nearly wrecked my car, and they haven't called you?"

"No, they haven't reported anything."

That's when I knew, right then and there: *I'd been had.*

They insisted on taking me to the hospital, which meant the cops would be left to find the dudes who shot me and to track down what happened at that burger joint on their own. I wouldn't be allowed to participate in the investigation anyway, but there was no question in my mind that the whole thing was a setup. They were just waiting for the right victim to come along. I had no doubt that their plan was to jack my car. What else they had in mind for me, I have no idea.

I'm sorry, but there was no way anybody was stealing my Mustang.

I thought the whole thing through, over and over again, trying to remember every detail of what happened. I was lying in the hospital when my buddy Scott came running in. This was just two or three weeks after we'd buried our close friend Ted, and Scott was a complete wreck.

"Dude, I'm fine! I'm fine!" I said. I couldn't help but laugh. Poor Scott thought he'd lost another friend.

The fact is, I was lucky. Twice in a month, I'd been lucky.

The doctors told me I'd been shot with a .38, and the cops who investigated found multiple 9mm shells at the scene from the second gun. Through some miracle, the .38 slug didn't hit any bone. It hit my muscle mass and left a long trail of a scar that's still visible plain as day today from the point of entry right down to where it ripped through the inside of my arm near my elbow.

My Mustang didn't suffer any bullet holes. Just me.

That bullet could have blown the hell out of my arm. Just a couple of inches over to the right and it might have gone through my back and into my chest, and I'd be dead. But to me, the moral of the story is that if I'd had my gun with me, I wouldn't be here right now at all. One of those guys might be dead, too, but grabbing my gun would have caused enough hesitation that I would have stayed there one more second, and I am absolutely sure that a second bullet would have come my way. Instead of hitting the gas pedal and peeling out from that bent-over position, I might have sat in place for another second—and that one extra second might have meant the end of me.

My 1965 Mustang 2+2 fastback. *COURTESY OF RICHARD RAWLINGS.*

I tell people that story when they talk about buying a gun to defend themselves, or keeping a gun in their house to protect their family. It's just something to think about. Sometimes having a gun can get you into more trouble than not. Sometimes it can cost you your life.

It took a lot of therapy to build the strength back up in my left hand. I also wound up being out of work for a long time, not only because I had to let the wounds heal, but because the department was scared about post-traumatic stress disorder, or that the experience might turn me into a bad cop who'd be out for revenge. I had to go through all kinds of psychological, medical, and physical testing to get back on duty. It took close to ten months.

In the meantime, everyone I knew kept telling me I should go after the burger joint. They were part of a chain of restaurants, and a corporation could be held accountable for the actions or inactions of their employees, I was told. At the very least, I knew that no one from that restaurant called that shooting in to the police.

The thing is, I'm not one of those people who's out to sue everybody. I hate that we're a country full of idiots filing idiotic lawsuits. So I ignored everybody's urging. I ignored it until about six months later when my roommate was in another nearby restaurant and he overheard two ladies behind him talking about how they'd been beat up and had their purses stolen inside the women's restroom at the same burger joint. "Hey, my buddy got shot there!" he said to them, and those women were like, "Oh, there's so-and-so who had this happen to him, and this happened to so-and-so." Hearing those stories made me angry.

I hired a lawyer and we pulled the restaurant's records and receipts. It turned out that girl at the window didn't even ring up my order. Which means at the end of the day, I got shot for twenty bucks.

The lawsuit dragged on and on. The attorney I'd hired said it could be years before I'd see any kind of resolution. I couldn't see that far into the future at that point in my life. I didn't have that kind of patience. Getting shot has a way of changing your thinking, too. You get pretty anxious about what you're doing with your life when you're confronted with the possibility of your own death.

A few older guys who I rode Harleys with had been trying to get me to sell them the Promo Wipes business around that time, and I said, "Why not?" I didn't see myself staying in the towel business my whole life, you know what I mean? So I sold it and made a few bucks. It was right around this same time when I sat down with my fire captain to talk about my future. I got talking to him about pay scales and how I could make mine rise a little faster, and he sat there and showed me how if I stuck it out, I'd be making X amount of dollars in seventeen years, just like he was making.

"Wait a second," I said. "No disrespect, sir, but after seventeen years, your pay has only gone up by that much?"

"Well, it's a lot more than you're making right now," he told me. "And when you consider the retirement benefits, you're on track to have quite a payout and retire real young!"

I thought about the other guys in that department who were basically waiting to retire—the guys who'd been so negative about my dreams and ideas—and all I could think was, *Is that what I'm gonna be someday?*

The fire department didn't offer me any opportunity to get ahead in life except to stay the course and know I'd have X amount of dollars at the end of so many years. It seemed way too finite for me. It felt almost claustrophobic or something. I'd joined the fire department because I wanted to get ahead in life. Suddenly, being in that job felt like the opposite. It felt as if staying would amount to nothing more than settling—and settling was something I never wanted to do in my life. Ever.

I stood up right at the captain's desk in that minute and said, "I quit."

"What? What do you mean, you quit?" he said. He thought I was just being rash and would change my mind, but I knew right then and there I wouldn't. My career in public service and safety was over. Life's too short. I wanted to make something big for myself, and I wanted to do it my own way. I'd started and sold the Promo Wipes business. I was sure I could do it again. It turned out that I was a heck of a salesman, and sales always offers a chance to decide your own fate and paint your own destiny.

"Nope. I mean it, sir," I told my captain. "Thank you very much for your time and all your help, but I'm gonna go do something else now."

I was barely going on twenty-five at the time, and I'd already had a career path and built a company of my own. I'd already gotten married and divorced, too. I almost forgot to mention that. I'd got involved in one of those tumultuous relationships where we broke up and got back together all the time, until finally it was like, "Are we gonna break up for good or get married?" And we went out and got hitched. It was a big mistake from the start, and I wound up moving out and living in my own apartment again not long before I quit the fire department.

In a way, I think I might have been moving a little too fast for my own good in those first few years out of high school. I'd maybe squeezed a little too much living into too short of a time frame. I needed a break. I needed to slow down. I needed to do that great American thing that I hadn't done yet: I needed to hit the road and go "find myself."

Man, I hate that term. I really do. I mean, I knew who I was. I knew who I wanted to be. I just wasn't sure what was going to come next, and I needed to break free of the expectations of everybody else around me before I could get there.

So what's a guy to do when he's feeling done with everything in life?

First of all, if you're me, you go buy a new vehicle. I'd sold the Mustang and traded for some other classic I can't even remember before I made my decision to up and leave the life I knew. I also knew I needed a really reliable vehicle if I was planning to go drive off on a great adventure. So I traded whatever car I had, and I used the good credit I'd built up as a fireman (before I let anybody know I'd quit) to run right out to one of the big dealerships in town and buy myself a brand-new Jeep Wrangler—the kind with the fold-down top. What better vehicle to go out and see America in, am I right?

What else does a guy in his mid-twenties do when he wants to change everything? When you're me, you sell everything. Just dump it. I took out an ad in the newspaper and I put some signs up in front of my apartment that said, EVERYTHING MUST GO!

I packed a few T-shirts and jeans and a cheap leather jacket into the Jeep that morning, and then I sat there drinking beer while people came and went from my place. I was absolutely determined to sell everything. People would look at the couch or a table or something and say, "Is this for sale, too?" I'd answer, "Everything's for sale. And anything that hasn't sold by five o'clock is going out on the front lawn and you can have it for free."

That's exactly what I did, too. At five o'clock that day, I threw the rest of my stuff on the lawn and I drove over to see my dad to say good-bye. I didn't know where I was going. I figured I'd head west, maybe to California, because that's just what people do. I had no idea when I'd be back, though, and no idea how I'd pay for anything once I ran out of the $5,000 or so I had in my savings account at that time.

I think that good-bye moment was the only time I'd ever seen my dad cry. He was scared to death. He thought he was losing me. He probably thought I'd lost my mind! Here I'd been this incredibly dedicated, hardworking, successful, clean-cut kid from the moment I left high school, and he just couldn't understand what in the hell I was doing.

"I'll be fine, Dad," I told him. "I just need to do this."

ON THE ROAD

Me, in the middle of growing up . . . and growing out my hair. *COURTESY OF RICHARD RAWLINGS.*

I didn't have any tattoos when I left home. There wasn't a single hole in my body that I wasn't born with—except for the bullet hole. My hair was so conservative it was almost military-style short. I barely even recognize me when I look back on those days, and you can be darned sure nobody who's a fan of *Fast N' Loud* would recognize a picture of me at twenty or twenty-two years old if they saw it anywhere outside of this book.

Anyway, I hopped in my brand-new Jeep and took off, determined to see California, excited to see the Pacific Ocean, and thinking that somehow hitting the road for a couple of weeks would make it all come together. I really didn't think I'd be gone for more than two or three weeks. The purpose of this trip was to figure it all out. What it actually turned into almost instantaneously was me camping out or staying at run-down motels, drinking beers at night, and having fun with whomever I happened to meet along the way. Once I was a few hours west of Fort Worth, I realized that I could basically do whatever the hell I wanted to do.

I was free!

I stopped for gas in Tucson and this backpacker dude—not a dirty, nasty-looking vagrant type, but somebody right about my age who just obviously lived on the road—came up to me and said, "Hey, can I borrow some money?"

"Dude, I'm strapped," I said. "I'm sorry, man, but I just don't have it."

"Well," he said, "could I get a ride?"

I tried to be nice and get out of it, but then finally I was like, "Sure. Where you going?" He said he was going east. I said I was going west. He said he didn't care, and he hopped in.

As we were cruising west and talking, it occurred to me that this guy somehow made enough money to live and eat and keep himself pretty happy, all while seeing the country, meeting new people, and moving from place to place. It intrigued me. I wanted to know how he did it. So I made a deal. "You can ride with me, and I'll buy you hamburgers and beer and what have you," I said, "but you've got to do something for me."

"What?" he asked.

"You got to teach me how you survive and do all of that with no job!"

He was pretty happy with that deal. So we talked for a bit, and I got a few little lessons from him on how to panhandle for money at the gas stations. A ways up the road, I parked my brand-new Jeep with the dealer tags on it around the back of some gas station, went out in front, and started asking people for money. Much to my surprise, people started giving it to me! An hour went by and I had $60 in my hand.

I'm thinking, *Holy s—t! This is insane!*

The last guy I asked gave me a $10 bill. I turned around, went right into that gas station's store, bought beer, beef jerky, and Cheetos, threw it in the Jeep, and we took off. The guy was still pumping his gas! I'm sure he was thinking, *What the f—k?*

It was so much fun that the backpacker dude and I wound up tooling all over the West Coast for months pulling this same routine, staying at dingy motels and gettin' loaded. We'd take turns driving so we were safe, but there were times when I'd just be sitting there drinking a beer in the passenger seat, watching America fly by, feeling freer than I'd ever felt in my life. We'd stop at whatever dive bar we could find and try to pick up girls. Sometimes we'd camp out in a public park somewhere.

It was like my personal version of living free like people did in the 1960s, and I had a blast. It was also a little bit like *Fear and Loathing in Las Vegas*, if you know what I mean. I did things I'd never imagined doing. We got crazy. And I started to look different, too. I let my hair grow for all those months. My skin got all dark and leathery from the sun and wind and partying. I had my backpacking friend pierce my ear with a straight pin on a park bench one night, and it's never closed up after all these years. I can still put an earring in that hole if I want to.

At some point we finally made it to Los Angeles. We were cruising down the Sunset Strip, right where the big billboard is near the infamous Chateau Marmont hotel, and I started looking at this gorgeous blonde chick in the car next to mine.

"Hey, how are you doing?" I said, and she answered me. She was hot, man! The sort of California-girls hot you hear about in old Beach Boys songs or something.

I said, "Hey, we should have a drink or something sometime," and she said, "Sure!" I told her I was new in town, and she said she was on her way to a meeting, but we made plans to meet at this certain bar after she was done. We did all that right there in our cars!

As she drove away, I looked at my trusty passenger and said, "Dude, this is where you get out."

"Cool," he said. He shook my hand. He grabbed his duffel out of the backseat, hopped out onto the sidewalk on Sunset Boulevard, and walked away.

It was the last time I ever saw him.

The blonde from the car actually followed through and showed up at the bar. I wound up staying on her couch for a while in L.A., where I completed the demolishing of my formerly clean-cut look by getting my first tattoo—a tribal band high up on my bicep, where it could be hidden under a short-sleeved shirt (in case I ever wanted to get a respectable job again, I figured). I actually wrote a check for that ink to Easy Riders Tattoo on Melrose.

Then I decided, Okay. It's time to go back home.

The thing I realized when I was out on the road is that I'd been the one pushing myself too hard from the start. It wasn't anybody else who was pressuring me into doing all the things I did. It was me. And I liked that pressure. I liked pushing hard to see what I could accomplish. I just hadn't given myself enough downtime. I didn't go to college and mess around for four years

A photo I keep in my office these days, which shows off my first tattoo. *COURTESY OF RICHARD RAWLINGS.*

like a lot of kids. I needed to blow off that steam. I needed to let go and find my own sense of style and everything else. It felt good, I had a blast, and when it was done, it was done.

"See ya!" I said to that blonde chicky and the other friends I'd made in L.A.

I hopped in my Jeep with the dealer plates still on it and headed back toward Texas. We didn't have cell phones back then, and I'd pretty much neglected to stop by a pay phone in all the months I was on the road, so nobody knew where I was. Not my dad, not Daphne, nobody.

I stopped in Las Vegas on the first night of my drive home and checked into the Imperial Hotel. Vegas seemed like a pretty rad place to be, so I decided to stay for a couple of nights—but only a day or so into it, I realized I was out of money. The check for the tattoo wiped me out. I was flat broke. No backup. No savings. Done.

That's the first time I called my family.

"Hey, I'm in Vegas and I'm out of money, and I need you to send me some cash to get home," I said.

My dad was pissed. He was like, "You got yourself in it, you can get yourself out, man!"

I didn't expect that response from him, but he was right. It took me about a week or so to scrounge up enough to pay for the hotel and to gather enough gas and food money to get back home, but I did it. I wound up at Daphne's house, where I crashed on her couch. They all looked at me like I was some kind of a street person or something. No one could believe how different I looked, and how different I seemed. That trip was definitely a trip, I'll tell you that.

Once the journey was over, though, reality hit and hit hard.

While I was lying there on my sister's couch on my very first full day back, some repo man showed up and repossessed my Jeep. I hadn't made a single payment on it since I'd driven off the lot. *Whoops.* So not only was I broke, but now I didn't have a ride, and my credit was shot on top of it all.

My sister and my dad were really worried about me, and I can understand why, but they had no idea how driven I was. I knew at that moment that no matter how down I got, I would never give up. I'd gone through my personal cross-country odyssey. I'd had some crazy, crazy times—many of which I couldn't remember if I tried to—but now I was ready to get to work. I was ready to take control of my own destiny.

From my experience with the Promo Wipes, combined with my ability to talk people into giving me money on the road, I came to the conclusion that sales was the first thing that I ought to jump into in order to start making some money real quick. I started looking through the paper and came across a job listing from a brand-new company with a brand-new concept. I called 'em up and discovered that this company was looking for a salesperson to go around to all the local bars and restaurants and talk to the owners and managers about us paying *them* to allow us to hang advertisements over the urinals in their bathrooms.

I was like, "Really?" and the owner was like, "Yeah. We've got another team selling all the big ads to the liquor companies and tobacco companies and all that, so it's up to you to go around and pay the restaurants and bars to allow us to hang those ads." I guess the liquor and tobacco companies figured they'd have a rapt audience while guys were standing there taking a leak. The concept seemed cool to me from a marketing standpoint. It was kinda genius, actually. I wished I'd thought of it.

"Sign me up!" I said. And so my first sales job was literally selling ads over urinals. It was a little bit embarrassing, and a little bit intimidating, but it allowed me to cut my sales teeth very quickly. I was thrown right into the pits where the only money I'd make were the commissions on the ad space I secured. And I nailed it. All of a sudden I was making some money. It wasn't great money, but it was money—and the more ads I sold, the more money I made. I liked that concept a lot. I liked being in charge of my own destiny. It felt right.

I said good-bye to Daphne's couch and went out and got myself another apartment. My credit wasn't great because of the Jeep repo, but I managed to fill my new place up with brand-new furniture and everything I needed by using store credit. It cost me a lot of money every month, and I racked up a lot of debt in a short amount of time paying insanely high interest rates, but I knew that as long as I kept selling ads, I could always make ends meet.

I guess I developed a penchant for living slightly beyond my means pretty early on. It kept me hungry. It gave me something to chase. It kept me on my toes. I like that feeling of knowing I need to get up every morning and go get it in order to keep living the life I want to live. It keeps a certain fire under your feet, and that's a good thing.

I was good at my job. There's hardly a better feeling than knowing you're good at something, and then going out and nailing it every day. That gave me confidence in every part of my life—especially with women. My edgier look attracted a whole different class of chicks, and walking around feeling like I had enough bank to do whatever I wanted didn't seem to hurt my chances, either.

I'd been at that sales job for maybe a year when I finally decided to take my confidence to another level: I decided to grab that long-lingering burger-joint lawsuit by the balls.

I was sick and tired of waiting around for my lawyer to get stuff done, and I was sick and tired of waiting for that company to pay me something for what I believed was their role in my getting shot. The whole thing had dragged on for almost three years without a resolution, which is the problem with lawsuits and one reason I am not a big supporter of engaging in any type of legal action unless it's a last resort. My attorney insisted we had a slam-dunk case, though, and that the only thing the burger joint had been doing was trying to stall and put us off.

Finally, the court put us into mediation, and that's when I put my salesman's negotiating skills to work.

The lady from the burger joint's corporate side walked into the mediation room and I said, "Look, I'm tired of this s—t. Bottom line is you know you are f—ked if this goes to court. Your job is to not pay me, or at the very least to pay me as little as possible. So tell me this: what's your deductible?"

She looked a little stunned.

"What's the company's deductible on their insurance?" I asked.

She told me that it was a hundred grand, which sounded like a large sum of money for me. I'd never had any money like that. I said, "Well, how about you cut off a little bit from that, so you can say you did your job—why don't we make it like ninety-seven thousand or something like that, and we'll wrap this up right now. You're gonna have to pay that anyways, so why not get this over with?"

"Done," she said, and they wrote me a check.

That was the first time I'd ever had any substantial amount of money in my life. Prior to that, no matter how hard I'd worked, I'd never had more than five or ten grand to my name at one time. It's kind of crazy that it took a bullet in my shoulder to get me that kind of financial payout. But such is life. Depositing that check made me feel like I'd won the lottery. I thought everything was going to be great from that moment on.

Of course, that's when I went and did one of the stupidest things I've ever done.

I had been in the process of building my credit back, and because I'd been making all of my payments on time, my credit score was actually pretty stellar. I was sick and tired of paying all kinds of payments every month, though, so I sat down on the very day the check cleared and did the responsible thing: I paid off all of my debts. I sat at my kitchen table and wrote out $38,000 worth of checks for all the furniture and credit cards and everything. It felt so good to be zeroed out and debt-free that I called up my credit card companies and canceled all of my cards right then and there. *Boom!*

I am never going to carry any debt again! I told myself.

Then I rolled out to the Harley dealership and bought me a brand-new Harley. My first ever brand-new Harley! Cash. Life was good.

By the time I was done paying taxes, that whole $97,000 was gone. I didn't think anything of it because I was debt-free and feeling good riding around on my new bike. Of course, a few months later, the ad sales started slowing down. The money got tight. I realized I needed a new refrigerator or something and that I didn't have the cash in the bank to go get it. So I filled out a credit application—and I got turned down. I'd never been denied credit in my life. I'd worked hard. I'd been a firefighter and a cop.

I'd always received great treatment from the credit companies. I was like, "What happened?"

Turns out that my credit was always good because I carried a certain amount of debt. Once I closed out all my credit cards, thinking I was being responsible, all that was left on my credit report was the repoed Jeep! It looked like I had no other actively *good* credit whatsoever, even though I was debt-free. I'd been stupid. I didn't know how the credit-reporting system worked before I made that bold move of canceling my credit cards, and I screwed myself over by acting rashly.

It would take me a solid five years or so to rebuild my credit after that moment.

Since I didn't have any credit anymore, and I didn't have the cash to simply buy the things I wanted or needed to buy, I realized I needed to find a new job that would give me a much bigger upside in sales commissions than the over-the-urinal ad business could provide. I couldn't find another job quickly enough to get me over the hump, though, and suddenly I was broke. I wound up having to sell my new Harley just to make ends meet, and of course I took a bath on it since any new car or bike depreciates like crazy the minute you drive it off the lot.

Still, I didn't let it deter me. I knew I'd never quit. I knew it was just a bump in the road, another lesson, a slap in the face to tell me I needed to work harder, and smarter, and to truly live up to my potential. Selling bathroom ads was clearly not my life's ambition.

As I watched some other dude ride off on my Harley, I vowed that I'd save up my money and buy myself another one—in cash—within three years.

I'd wind up doing it in six months.

OVERDRIVE

Funny thing about setting goals: sometimes once you set 'em, you exceed 'em.

I started reading all sorts of motivational books to help me with my sales career, and I've gotta tell you, that's a trip in and of itself. Everybody should read those kinds of books at some point in their life. Reading about the road to success and the path to greatness can get you all sorts of pumped up about your life and possibilities. The key to applying the philosophy in any of those books, though, is to realize there's no shortcut. There's no gimmick. Big success takes huge amounts of work. You just need to hold on to the enthusiasm and passion for whatever it is you're doing, and then go do it. Anything short of getting it done is just laziness in my book. It's all up to you. There's nobody holding you back but yourself. Ever.

I gave myself three years to save up enough to buy myself a brand-new Harley, sure, but I also knew that the Harley was just one small goal of mine. In the back of my mind I knew that I wanted to be able to get back into cars, and not just the way I was into cars in high school, buying and selling 'em for a few thousand bucks, but big-time into cars—buying and selling classics, going to the big auto shows, going to car auctions, building hot rods, eventually buying Porsches or Ferraris or Lamborghinis the way rich folks did and driving like a bat out of hell just for kicks. I wanted to be able to buy nice things in general. I wanted to live the life of the rich and famous. Basically, I wanted to live large, all while having as much fun as I possibly could.

I know they say money doesn't buy happiness, but damned if you don't need a whole lotta money in order to buy things that can bring you the kind of life I wanted to live. And living the life you want to live is what brings you happiness, isn't it? That meant I needed to make a lot of money, and to do it before I got too old to enjoy it.

In my search for a new sales job that could make me a lot of money, I ended up meeting a chick who worked in the printing business. This girl was a saleswoman, and she was successful at it. She was pulling in like $50,000 a year, which was a pretty darned good living in 1996. I was intrigued by that. I knew a little bit about the printing and paper business already because of my earlier success with the Promo Wipes business. I also knew a thing or two about sales, and definitely had a natural knack for it. So I asked this girl if she could introduce me to her boss, and she did, and next thing I knew I was working as a salesman for a print broker.

Basically, this guy was the go-between who would broker deals between companies that needed something printed—be it business cards or menus for restaurants, or giant billboards for some big company—and the design firms and print houses that could get the work done. Printing was a much more complicated business back then than it is now. Nobody had access to computerized design programs and that sort of thing, so a broker could step in and provide a business with access to everything they needed, which would save them a lot of time. And in business, time is money. The premium a business would pay to have us take care of everything they needed is where the profits came in, and the profits could be pretty huge.

I made my money taking a commission from those sales, and I frickin' nailed it. I completely hit my stride in that salesman role, and in a matter of months, I surpassed that girl's sales for the year. I convinced the boss to give me a larger commission percentage if I could meet certain goals—big goals that he thought were impossible for one salesperson to meet—and then I went out and I hit and exceeded those goals. I threw myself into making money the same way I'd thrown myself into goofing off and traveling around in my Jeep. The way I saw it, sales was fun. It was a challenge to get people to part with their money at the same time you were treating 'em right and making sure they'd come back for more. After all, making good money meant I could have the things I wanted.

Six months into this, I cashed a single commission check that was more than enough money to cover the cost of a brand-new Harley. Not just any Harley, either. At the tail end of 1996, I picked up a '97 Harley Davidson Softail chopper in my favorite color, green. I got that chopper custom painted with all sorts of crazy skeletons and detailed scenes that you can only see when you're right up close to it, and I still own that bike to this day. I love it. I promised myself I'd never get rid of it. I keep it right outside of my office at Gas Monkey Garage alongside a bunch of my other personal cars and bikes that I've collected over the years.

I cleared over $100,000 in my first year in that job. I also fell head over heels for the girl who got me into the business—a relationship that would affect my life in some very significant ways, and also one that wouldn't end well. But that's a different story for a different book.

Once the boss saw how well I was doing, he very quickly made me VP of Sales and gave me the leverage to make some changes in the company. I encouraged him to buy some printing equipment himself, so we could sort of double-dip and make profits both from the sales and from the actual printing. I found ways to incentivize the other salespeople to go out and hustle up more business the way I did, too, all while increasing our profit margins and putting more money in their pockets. I threw everything I had into that business, and every idea I had worked.

I never let my dedication get in the way of my fun, though. I'd take all that cash I was making and go hit the road on my Harley on the weekends. I occasionally took crazy cross-country vacation rides with some fellow Harley enthusiasts, blowing money on beer and women everywhere we went.

By the time I was twenty-eight, I was pulling in somewhere around $400,000 a year as a salesman. I was driving around in the biggest, baddest Cadillacs I could find. I had a brick phone—you know, those big old cellular phones with the antenna on top—before anybody else in the Dallas area had a cell phone. I could afford just about anything I wanted, and what I realized was that everything I had still wasn't enough for me. I wanted more. I wanted everything.

It was right around that time when I met a girl named Sue.

Sue was the prettiest tall, skinny, buxom blonde that I'd ever laid eyes on—and I'd laid eyes on a lot of tall, skinny, buxom blondes. (I most definitely have a type!) The thing about Sue that very quickly blew me away is she was easily as smart as she was beautiful. She was also extremely successful. She was running a multimillion-dollar home-health-care company. She drove a BMW 7 Series. She had a beautiful house with a pool in a gated community in the suburbs. She didn't need a man to support her. She didn't *want* a man to support her. And I gotta tell you, that right there is one very sexy quality for a woman to have.

This girl rocked my world so hard that on our twenty-eighth day together, as we were driving past the airport on the way to the city of Grapevine for dinner, I looked at her and I said, "Right or left?"

She seemed confused. "What?"

I said, "Well, we could go right and go to dinner like we planned, or we could go to the airport and fly to Vegas and get married instead."

"Are you serious?" she asked me.

I was sort of half joking when it came out of my mouth, but all of a sudden, it hit me: I really wanted to marry her!

"F—k yeah," I said. "I'll go to Vegas. Let's go. We'll get married."

She said, "Left!" So we drove into the airport, went up to the ticket counter, bought two tickets to Las Vegas, and got hitched.

We hardly spent a day apart after that, for many years.

With Sue's encouragement, I broke free of the job I was in, took my clients with me, and started my own print-brokering business. The amazing thing about that business is that all I needed to run it was a desk and a phone. There was very little overhead.

I saw changes coming in the print business, though. Computers were changing everything. The old typesetting services and glass plates and all the things that had dominated the field for so long were simply going by the wayside. Working around it all the time, I could see that what businesses really needed was one-stop shopping: companies and individuals needed to be able to walk into one place to have their designs done, and their printing done, and to have all of it managed under one roof without brokers or anybody else in the middle.

I was lying there in bed one night with Sue when I told her what I'd been thinking. "Sue, I want to do something silly," I said. "The broker business is going to die because things are getting more computerized. So in order for me to keep my clients, I need to buy the printing presses to do most of the work." I also told her I was thinking one step further. "I think the ad-agency mentality is going to go away, too, especially for the small-business owner, the medium-business owner, because the computers are getting smarter." I said, "I want to have an art department that can design and do brochures and menus and packaging and all of it, and then print everything in-house. That equipment, it's expensive, but I want to bring it all under one roof. We bring in everything from the very beginning—design and art to proofs to prepress to our own press, press checks, whatever we want—and we do it all right there."

She thought I was crazy at first. But the more I talked about it, the more she saw my passion and vision. I was positive I could make a profit on every single aspect of the business. Finally Sue said okay, and with the help of her reputation and good name, I got a small-business loan. I bought an established printing business in town so I wouldn't have to start from scratch, and then I started to build my own little company.

I know what you're probably asking at this point: what in the hell does any of this have to do with building hot rods, flipping cars, and the craziness that we see every week on *Fast N' Loud*? Well, let me tell you: business is business. Looking at how to do things faster, smarter, and more profitably than the next guy is important no matter what business you're in. The lessons I learned in the print business would feed absolutely everything I would eventually do at Gas Monkey Garage.

I put in a prepress, put in an art department, put in design. I hired all those people. Then I built my salespeople up, too. Here's where I got smart: I treated all my salespeople as brokers and double-dipped on the profits. I priced all of my print jobs at a profit to begin with, and then my brokers would go out and sell our services for a markup from there, and I would pay them well for doing that. They were basically allowed to set their own pricing, and they would make a commission of 40 percent of whatever profit they delivered. Of course, that meant I was keeping 60 percent of that additional profit for myself. *Boom!* It was rad. In no time at all we were kicking ass and taking names.

I also put a premium on customer service. I knew from watching my dad work in the grocery store that if you treat a customer right, they'll remember you, and they'll come back. Most print shops in those days were grimy, dirty places where nobody would ever want to hang out for longer than they had to. I built a shop that was sparkling clean from top to bottom. If a customer came in to check his or her proofs, there was a glass of champagne waiting for them upon arrival. If a customer came in to sit on press checks, there was pool table and a little lounge area and a bar and a beer cooler for their enjoyment while they waited.

We quickly gained a reputation as the best medium-sized print shop in the Dallas metro area. Before I knew it, I had $2 million worth of equipment and a great big staff humming away under one big roof, just like I'd envisioned.

What's really amazing to me is that I hadn't laid out all that much cash to get it started. I'd followed my shop teacher's advice and managed to sock away some savings while I was working for somebody else, and I'd risked a huge portion of that meager savings to get this new company off the ground. You've got to take the risks if you want to get the gains. But small-business loans were what paid for the building and all of the equipment. The cash layout was tiny compared to what the company grew into. There's a whole world of possibility out there if you look for it. As long as you're smart about it and you put in the hard work, there's no reason you can't start a business with very little cash of your own. None of the businesses I started cost a fortune out of pocket at the front end. None of them! Including Gas Monkey! But I'm getting ahead of myself.

I hired my sister, Daphne, to come in and do all the books for me. Prior to that, she was working for another company that was actually one of my biggest customers. She was a huge asset in that she could share information with me about all of the other print shops in town and what they were making in terms of competitive offers—so that allowed me to always be a step ahead of them in my pricing. I had eyes and ears everywhere, actually. I was always watching what the competition was up to and doing my best to get in front of them in terms of aggressive pricing or new equipment or whatever I needed to do in order to win new clients and keep my old clients coming back for more.

I also refused to put all of my eggs in one basket. A lot of print companies relied on one big client for 40 or 50 percent of their business. If something happened and that one big client went away, the whole operation could go belly-up. I refused to let that happen. If any one client got to be more than 10 percent of my business, I would tell them to take a portion of their business elsewhere, or I'd just kick them out entirely and tell them we could no longer service their needs. I know that seems backward to some people. I'm not most people. Staying diversified kept my business healthy, and I knew it was the right way to run things because my business stayed profitable while I was paying my salespeople more than they would get paid at any other print shop around. I wasn't the only one rising as the business rose. I loved seeing the people who worked for me get more successful, too.

So scratch that earlier statement. My business wasn't just "profitable." My business kicked ass! And that left me room to have some fun.

Honey, pack your bags. We're going to Florida."

One of the best things about Sue was I could talk to her about anything, from business to cars, and she'd be into it. She not only supported me, she understood my passion for things. In fact, she was a car fan herself, and one day she told me that the '68 Shelby was her favorite car ever. She'd never owned one. They were rare, and expensive, and she'd just never come across one that was up for sale.

"What are you talking about?" she said that day when I called her.

"I just found a guy down in Florida who's selling a matching pair of '68 Shelbys. One for you, one for me. Let's go get 'em," I said.

The next day we were off to Florida, where we spent $65,000 on a matching pair of red Shelbys. Hers was an automatic GT 350, mine was a stick GT 500, and to see those two cars side by side was one of the coolest things ever. We drove 'em to the beach, sat in the sun, and had a few drinks, and came up with the bright idea to commemorate our new purchase with a set of matching tattoos. Together we drew up the back-to-back style "RR" that I got tattooed on my arm that day, and that she got tattooed on her back. That night, we went to one of those crazy beach-town bars where people write their names or some funny thing on a five-dollar bill and then staple it to the wall. The whole bar was covered with these bills, and so I decided to do it, too. I grabbed a pen and wrote a little poem: "Two Shelbys, two tattoos, one for me and one for you." I stapled it on the wall.

Back in Dallas, driving that Shelby around made me feel like I was on top of the world.

I wanted more of that.

A snapshot of one of our Shelbys. *COURTESY OF RICHARD RAWLINGS.*

At the same time I was building the business and buying nice cars for my wife and myself, I developed a close friendship with a guy named Dennis Collins. Dennis and I had sort of grown up together, even though we didn't know each other. We'd run in some of the same circles. We'd often wind up in the same bars at the same time. We were both into cars. We both rode Harleys. We were like two peas in a pod, just not in the same pod. Then we wound up taking a cross-country Harley trip down to Daytona with a bunch of other local guys, and we started talking and partying and whatever. He was a very cool guy, and I was glad to know him. But it wouldn't be until a couple of years later when we'd really connect.

In 2003, I decided to enter a road rally called the Gumball 3000. This was a three-thousand-mile race full of wacky guys in their fancy cars who were hell-bent on driving as fast as they could from San Francisco to Miami. I'd been reading about this thing since the first Gumball started in Europe in 2000, and one night I just up and decided I was gonna do it. Sue thought I was nuts to throw myself into that sort of thing, and so did a lot of other people. Other than taking my Jeep to California and riding my Harley, I didn't have any experience with driving cross-country, and I certainly didn't have experience driving in a non-sanctioned competitive road rally at high speeds. Oh, and I didn't have a fancy fast car, either. You know what I entered into that rally? My Chevy Avalanche pickup truck. I knew that truck could haul ass, and I also knew that I was more driven to complete and hopefully win this thing than just about anybody else in it. When I get my mind fixed on something, I hit it. Hard. And I certainly wasn't scared about it. To me, it just sounded like one of my cross-country road trips kicked up a notch. It seemed like a way to live out a *Cannonball Run*–style dream. I mean, is there any American guy my age who didn't grow up dreaming about driving some fast car across the country like they did in that movie? I'd built a business that allowed me the money and time to go live out that fantasy. Why the hell wouldn't I give it a try?

When I got out to San Francisco for the start of the race, who did I see in the car right in front of me? Dennis Collins! He'd entered the Gumball 3000, too!

From that point on, Dennis and I were thick as thieves, man.

Oh, and by the way: I won. My very first Gumball 3000. I beat the other guys by nearly two hours. In a Chevy Avalanche.

I'm no race-car driver. I don't even think my cop training helped me to become any kind of a better driver than anybody else. But I do have an intense amount of focus, and if anybody had any doubts about my stamina and determination when it came to driving, they were put to rest then and there.

Dennis and I wound up spending a lot more time together once we got back to Dallas. He'd grown a successful business selling Jeeps, and I very quickly learned what an incredible businessman he was. He saw things that other people didn't see, like the fact that the Jeep's body style rarely ever changes, which means there's endless opportunity for customization that never gets outdated. He capitalized on that in the biggest way possible and made a ton of money doing it, too. He was also into buying and selling cars just for fun. I'd managed to flip a few cars on the side during my twenties and early thirties, but never as many as I did back in high school or immediately after. I was hungering to get back into my car-flipping hobby, and Dennis and I discovered a whole new way to do that at the exact same time we became good friends.

EBay was a relatively new company in the early 2000s, and when they started allowing people to buy and sell cars on the Interwebs, a lot of people thought it was a crazy idea. It *was* kinda crazy. Just crazy enough for me and Dennis to have some fun messing around with it.

We started hanging out after hours at my print shop every Wednesday night for a little thing we called Wacky Wednesdays. Basically, we'd sit there drinking beers and laughing our asses off while we bought and sold cars on eBay.

Because people were just starting to adapt to the Interwebs, a lot of the people who were selling cars were still in the backwoods on what their cars were actually worth. We would steal some absolutely incredible bargains online, then flip them for a huge profit with almost no effort at all. I remember one time we snagged up a '63 Corvette for $15,000, knowing damn well we could sell it for $40K.

I picked up some hot rods here and there that I kept for myself, too. I put a couple of 'em on display in the print shop, in the same area where I kept the pool table and beer cooler for customers. People loved coming into the shop and sitting behind the wheel of those crazy old cars. It's hard to explain how much enjoyment people got out of it. They weren't even driving them, like I got to do. They were just sitting behind the wheel of some car they'd never seen before, in the lobby of my business, dreaming about what it would feel like to take it out on the open road.

Wacky Wednesdays became the highlight of my week just about every week. Of course Dennis's wife and my wife would get pissed that we'd come home hammered in the middle of the week, but we were both basically like, "Whatever." While we were drinking, it wasn't uncommon for us to make $20,000 in a single night. All while having a good time.

If that's not time well spent, I don't know what is.

By 2004, I saw that the printing business was about to go through another major change. Computing power was growing so fast, it became clear to me that the design end of the business was eventually going to be eaten up by every Tom, Dick, and Harry who thought he could make his own logos and layouts on some homespun version of Photoshop or something. The debt load my company was carrying on all of that equipment was starting to feel too heavy, too. Back when I got the loans for everything, interest rates were at 12, 13, even 14 percent. Interest rates in 2003 had dropped to 5 or 6 percent, but no one would refinance my equipment at that point. The company was so successful that I'd turned down a couple offers to buy it outright over the years. I was making too much money and having too much fun to even think about selling it. But a part of me was quietly starting to think that maybe I should get out while I was still on top.

That's when the owner of another major print shop in the Dallas area showed up at my office, completely out of the blue. He seemed to know all of the ins and outs of my business, right down to where I kept my financial notes on my desk. Clearly I wasn't the only one who had eyes and ears all over the place when it came to keeping tabs on the competition. One of my salesmen must've been ratting me out! Anyway, this guy sat there and ran through some numbers with me and asked me if they were true. I told him they were. He asked me if I really kept my business diversified to the point where no single client was responsible for more than 10 percent of my sales. I told him his facts were correct. Then he asked me what I realistically thought the business was worth, and I told him. I gave him a price. It wasn't a pie-in-the-sky price, but an actual price based on the actual numbers, which I ran in my head constantly and double-checked with Daphne every Monday morning like clockwork.

Right there, with no lawyers, no accountants, no nothin', that guy whipped out a pen and wrote me a check for the dollar figure I gave him.

I looked at the check and didn't even think twice. "Kick-ass!" I said.

I stood up, shook his hand, and I was done with the printing business. Just like that. I didn't make millions of dollars on that deal like some people think. That's just not the way a business like that works. Running a printing business, even a very successful printing business, is a way to make a living. I did it big, and I did it fast. I'm incredibly proud of what I built, and that shop is still very much alive under that new ownership. In fact, it's probably the biggest, most successful printer in the whole region now.

But that business wasn't big enough for me. I didn't want to "make a living." I wanted to make some serious bank. I wanted to build something bigger. I wanted to set the world on fire, you know?

That offer to buy me out came exactly at the right time, because I'd recently had a bit of a revelation. I finally knew what it was I wanted to do with my life. I finally knew how I was gonna reach my goal of living the fun, freewheeling, rich-and-famous lifestyle that I knew I wanted to live.

My true passion had been staring me right in the face all along.

NO MORE
MONKEYING AROUND

I could do that," I said.

"Do what?" my wife asked.

We were in bed watching TV one night shortly before I sold the print shop, and for some reason we happened to flip over to *American Chopper* on the Discovery Channel. I had seen the show before, and to be honest, I didn't really like watching it very much. It seemed to me to be nothing more than a bunch of people yelling at each other and fighting all the time. I got frustrated watching it. But I loved seeing the bikes they built and the work they did, and like a whole lot of other people who tuned in to that show, for some reason I couldn't turn it off.

I was aware of what Jesse James was doing over at West Coast Choppers around that time, too. His brand had blown up, and he was selling T-shirts and apparel all over the country, just because he'd managed to get himself on TV. As a businessman it blew me away to see how big all of those guys got, even though in my opinion there were other better bike shops and custom fabricators in other parts of the country doing work that was just as interesting. The only difference with Orange County Choppers and West Coast Choppers was the fact that they had personalities and brands being showcased on TV—personalities and brands that the general public clearly seemed to be drawn to.

All my dabbling in the car world had led me to learn a thing or two about Boyd Coddington, who was the premier hot-rod builder of his time and who'd lined up a TV show on TLC; and Chip Foose, who left Coddington to start his own company and was in the process of launching his own TV show, too.

"Hell, Sue. I could do it better than any of 'em!"

"What are you talking about?" she asked.

"I'm going to go after these guys. I think I can do it better. I think I can assemble a team that can build better."

Sue didn't doubt me. She just wasn't sure where I was going with all of it or why I was suddenly so fired up. "Well, yeah, so you could build a brand, but then what?"

"Then," I said, "I'm going to get on TV."

Sue just laughed. "Yeah, right," she said.

The next morning I sat down at the kitchen table and drew up a business plan. It wasn't anything elaborate. I just put a few thoughts on paper and got myself pumped about the possibilities of this new venture. No one had really focused on building cars on TV at that point. It was all about motorcycles. I did some reading about these shows on the Internet, and it seemed that the main reason people weren't building cars on TV is because there wasn't enough time. TV shows need to churn out episodes quickly, and motorcycles were just easier to build. You couldn't shoot one episode every six months. You had to turn out new episodes every week or two in order to make a TV show a success. I was sure I could assemble a team that could build cars just as quickly as those other guys were building motorcycles. Why not? Plus, I knew there was money to be made flipping cars. I had a pretty good idea that customizing and building hot rods for the types of guys who might have enough money to desire such a thing could be profitable, too. I had no doubt that I could go open up a garage and make a pretty good living doing something I loved—and with any luck, I could turn this thing into something much, much bigger than that.

I thought about how many guys are out there in the world with some cool car in their garage that they like to mess around with on weekends while they drink a few beers with their buddies. I thought about my own sense of fun, and how everything I did that was fun seemed to be tied to wheels. I used to say to people, "If we're gonna have fun, it better have a motor." I mean, doesn't everybody love cars at some level? Cars are about freedom! I really thought this idea could go somewhere. I just needed a name for it all.

Then it hit me: somewhere along the line I'd come up with a term for guys like me and those weekend tinkerers—guys and gals who loved nothing more than messing around with cars every chance they got. I called 'em gas monkeys.

Right there at that kitchen table, I decided that was it. My new business venture would be called Gas Monkey Garage.

I called up one of my former employees at the print shop and we designed a logo. It was nothing like the logo you see today, except for the fact that it looks sort of vintage. It featured a monkey skull and a checkered flag, in more of a skull-and-crossbones type of look. I'd completed a couple of road rallies at that point, and the idea of racing was strong in my mind.

I sold the print shop right in the middle of this crazy week of thinking and planning, and then a week later, Sue was out watering the flowers in the front when she heard the sputter-roar of an engine coming down our street. She turned and looked in complete disbelief as I came rolling into the driveway in a rusty old Model A open-top rat rod.

I was beaming when I hopped out. I was completely pumped! I'd noticed that there seemed to be a trend of people buying and selling more and more rat rods that whole year. Traditionally rat rods were these rusted-out, pieced-together, piece-of-s—t hot rods that people built in their backyards from whatever spare parts they could string together. But the newest trend was to keep the look of those old cars—the unpainted, rusty, rough-looking bodies— but turn them into really cool cars. Fast cars. Loud cars. Safe cars. Cars that could compete in rallies and races without falling to pieces whenever you hit a pothole. Cars with the best suspensions and brakes and air-conditioning and the works, all hidden under that hunk-of-junk look that made 'em cool. I bought that first Model A for $7,500.

"You sure as hell better know what you're doing," Sue said that day, shaking her head and thinking I'd lost my mind.

I knew exactly what I was doing, and I knew exactly the man I needed to go see in order to get my Gas Monkey dream off the ground.

I first met Aaron Kaufman a couple of years before this, mostly by chance.

I'd picked up a '51 Ford Customline that was in need of some sprucing up, so I took it over to a local auto body shop to see if they had anyone who might be able to do something with it. I wanted to put some new wheels on it, maybe add some cool suspension. They sent me straight to this twenty-year-old self-taught hot rodder named Aaron.

He did a whole bunch of work on it, and what really impressed me is that he went above and beyond. Every bit of work he did was methodically well done. He had to fabricate some components for it, because they just weren't available back then, and everything he did was super clean. It wasn't hack work like you get from a lot of shops that are just trying to turn a buck. Not only was everything where it was supposed to be, I could just tell that he cared about what he was doing. He took pride in it. He was the opposite of one of those mechanics who hates his job and smokes cigarettes all day and gets pissed off whenever a customer comes in. (Admit it: you know you've encountered those types of mechanics in your life, even if you *are* one of those guys!)

I told him I was impressed, I told him I'd be back, and a couple of months later, I was.

I've only been burned by a car I purchased on eBay maybe four or five times over the years, and one of the first ones that completely pissed me off was a '64 Galaxie convertible I picked up during one of my Wacky Wednesday sessions. Well, it was *supposed* to be a '64 Galaxie convertible, and that's certainly what it looked like in all of the pictures online. So I bought it, and I had it shipped in, and it turned out that it wasn't a real convertible at all. This guy had basically chopped the top off of it. It was nicely done, but it ruined the value of the car.

I took it to Aaron hoping he could do something—anything—to help me recover some of the money I'd spent once I went to resell this thing. As soon as I got there I started bitching and moaning about the money I'd spent on that car, and how I got ripped off, and instead of offering me suggestions on how to fix it, Aaron said, "Dude, I like the car. Why don't I just buy it from you?"

I think he took it off my hands for $4,500, and that was that.

A couple of months later I stopped by and Aaron showed me what he'd done to the Galaxie—and it was rad! He put twenty-two-inch wheels on it. Then he put a full air ride on it and laid it on the ground. That was cutting edge technology at that time. No one was doing what he'd done to that car. Frankly, I was upset that I'd sold it to him, 'cause I knew as I stood there that I could've turned around and sold that car for a heck of a profit if it was still mine!

Aaron just seemed to see things that other people didn't see. He was self-taught, and because of that, he saw possibilities in vehicles that were so far out of the box that the box wasn't even visible from where he was standing. I was impressed as hell. And that impression stayed with me.

When I thought of what I wanted to do with Gas Monkey Garage, Aaron was the first guy I thought of. I knew I needed to get something going before I'd be able to steal him away from the shop where he was already working, though, so I went out and rented a garage space. I hired some kid just out of college who'd be able to help with the basic flips, who could take care of putting a little shine on some old cars so we could turn them around and sell them. But what I really needed was a fabricator, a guy with some vision, a guy who could help me build a reputation for building great cars, fast cars, loud cars that would grab attention wherever they went. Aaron Kaufman was the guy I needed.

Finally, with a few things in place, I went to see him. "I'm going to start this company called Gas Monkey Garage. Here's the logo," I said, and I showed it to him. It was a really cool logo. It wasn't just black and white. It had some color. It had some attitude, you know. I tried to convey to him that the logo represented the attitude I wanted the shop to have. "So I got this little shop, and I want to build some cool cars, fix them up, sell them. And I'm going to buy a big rig of some sort, you know, a big truck and go out on the road to all the big car shows and rallies and sell T-shirts, and take some of our cars with us and try to show the world who we are and make some money doing it."

Aaron didn't seem real interested in the business aspect of it. He got pretty fired up about the car side, though, and the idea that he'd basically be in charge of his own shop.

"You really think you can do all that?" Aaron asked me, and I said, "Yeah. Don't you?" And he goes, "No!"

I was a little taken aback.

"Wow. You don't?" I said.

"No," he said. "I don't think you can, but I'm along for the ride."

We shook hands, and I hired Aaron right there on the spot. I told him I could pay him what he was making at his current job, even though I didn't know how in the heck I was going to do that after my funds ran out in a month or so, but I had a vision, and I had a plan, and I went for it.

"Cool!" I said to him. "Let's get to work!"

About a week later I went down to a hot-rod show that happened to be in Austin, just to check out the scene and see where Gas Monkey would fit in as I looked ahead to the coming months.

That's where I spotted the next big thing I needed to build my business: a big, full-on NASCAR-style, eighteen-wheeler-sized motor-home-and-trailer combo. The trailer held two cars up top and one below, with a full shop, all air-conditioned. The motor home that pulled that trailer had full living quarters and a shower, and slide-outs for extra space when you parked. It looked like a million dollars' worth of vehicle to me, truly like something an established race team would own. I looked at Sue and I said, "This is what I've been looking for."

"What?" she said. (She said that a lot. A lot of people wind up saying that when I get talking about my crazy ideas.)

"Yup. I'm gonna buy that rig, and I'm gonna put 'Gas Monkey' on the side of it, and I'm going to show up at these hot-rod shows and blow people's minds. Look around! Everybody else is here with crappy trailers and cheap flyers. They're barely making a living. They're not making money. This will draw so much attention, everyone and their brother's gonna wanna know who in the heck Gas Monkey Garage is and where they came from!"

"That's just crazy," she said.

"Well, I'm doing it. I want this rig," I said. I knew I didn't have the cash for that sort of thing. It was worth hundreds of thousands of dollars. But I went over and talked to the guy to ask him how I could buy it with very little money down.

The guy wasn't put off at all by my questions. He didn't think I was crazy. He may have doubted that I was serious about buying it, but he answered me seriously, which I appreciated.

"Well, I suppose you could finance the back as a trailer, and you can finance the front as an RV, and you know, if your credit's good, we can get you locked into that. The thing about this type of rig is that you can finance it for thirty years, you know, like a second mortgage."

The big Gas Monkey Garage rig! *COURTESY OF RICHARD RAWLINGS.*

I quickly did all the numbers in my head and realized I could own that unbelievably rad rig for less than $2,000 a month.

"I'll take it!" I said.

The guy said that one was just for show and wasn't for sale at that time, but he promised he'd have somebody call me. I gave him my number. Then nothing. I don't know if he looked at my long hair and tattoos and thought I was full of it or what, but a few days later, I tracked down the company, called the owner, and said, "Look. Here's the paperwork. I downloaded it off the website. It's all filled out." I said, "I'll be down there in a couple of hours and I want my damned truck!" And he was like, "Well, okay."

Sue and I drove down there and loaded her BMW right up into the back of the rig for the ride home. I'd never driven anything like that in my life. It was huge! The best part about it, though, was that I didn't need a truck driver's license to haul it. The size just barely fit the specs required by law for a recreational vehicle and trailer, so I didn't need to get a special license or special registration, I didn't need to keep log books or go through the weigh stations that commercial drivers face. None of it.

I maneuvered that rig pretty well, and just about blew Aaron's mind when I pulled it up in front of the shop.

"You ready to hit the road?" I said.

Aaron just shrugged his shoulders, shook his head, and said, "F—k it, man. Let's go!"

GEARING UP

Aaron and I spent the next few years driving to every car show and rally we could get into, all over the country, in addition to doing everything we could at the shop to get Gas Monkey off the ground.

I stickered our rig up, basically wrapping the whole thing in our logo, and honestly, it was bigger than a damn billboard. It was insane how big that rig was. I remember thinking to myself, *I've got to have something that makes these people believe that we are who we say we are—something that really makes an undeniable statement of "Hello, world! Here we are!"* That's Gas Monkey style all the way: kicking the doors in wherever we go, and not just showing up like everybody else. I guess that's how I got ahead in sales long before Gas Monkey came along, too. In a way, it really has been my whole way of doing life, period. If I'm gonna do something, I do it bigger and better than anyone else.

I pulled that rig into our first show in the Cow Palace in San Francisco, and Chip Foose was out there with just a regular pickup truck and a trailer. Boyd Coddington was there with something slightly bigger, but nothing eye-catching. And then Aaron and I come driving in in this gigantic rig that barely fits through the industrial doors of the Palace. All sorts of fans of those other guys were already there when we arrived, and you could see them all turn and stand back and stare at us like, "What the f—k?" People flocked to us just to find out who the hell we were, and we started throwing out T-shirts and full-color flyers. It was rad.

We started making some waves back home at the garage, too. From the start, I planned to build a couple of cars all on our own to show off what we could do. A lot of shops get so caught up doing work for customers that they're never able to turn around their own product, which can wind up being the stuff that sells for big bucks. Long-term builds for customers can also take up your entire shop, and then jobs start to pack up against each other and you get all backed up, and suddenly your customers are waiting not just months but years for their cars to get done. I insisted that would never happen at Gas Monkey Garage. We'd never get involved with a build that would take more than ninety days. So our customers would never be left in the lurch, the only work that we'd allow to get backed up was our own long-term builds. If somebody wanted something done, I'd guarantee them a completion date or they'd get their deposit back. Nobody else was doing that. It blew people away. Of course, I asked for more money in order to guarantee a completion date, but guys who are throwing money into hot rods immediately knew that it was money well spent. No one wants to buy a car and then have it sit in pieces in some garage for two years. You want to get out there and drive it! So that's what Gas Monkey guaranteed they'd be able to do, faster than any other garage could offer, by far. It wasn't rocket science. It was just a different business model. I was sure it would work. And it did. By flipping cars to make some cash here and there, and taking on projects that yielded results, I was able to pay Aaron his salary, pay the rent, and pay the entry fees we needed to participate in the hot-rod shows and big rallies that would build up some street cred and let people everywhere know what Gas Monkey Garage was all about.

Just as we got the shop off the ground, I realized there was another Gumball 3000 coming up—and I decided to run it not only for the fun factor, but for the promotional value. My wife didn't understand how going off to run a road rally at the same time I was starting a new business made any sense. So I explained it to her this way: "I'm going to be playing with the richest guys in the world. And I'm doing it as Gas Monkey this time. I've got a brand. I've got a shop. So this is going to get me clients. Plus, if I can win, if I can push them, then that's notoriety. That's in the newspapers."

I went out. I ran it. I won it, again. Suddenly Gas Monkey was on the map—and so was I. The car world is a pretty small one, and those road rallies and things don't get a whole lot of press outside of hot-rod and supercar circles. But I was on my way.

I was so fired up, I started doing all the road rallies I could find—and kept whupping everybody's ass along the way. There were times that Aaron and I were in the truck for two, three months at a time. We were crisscrossing the country, so there wasn't any time to come home. Along the way, we would pull into gas stations and throw out T-shirts and just spread the word, like we were on some kind of a victory tour. When we finally got back, Aaron got to work building the most unbelievable rat rod anyone had ever seen. My plan was to run it in the Bullrun—another big underground road rally with a huge following. The Bullrun, like Gumball, runs a route that's about three thousand miles, but it changes every year, and it runs through both Canada and the U.S. It also features a lot of exotic cars that have been super modified. We're talking all kinds of top-of-the-line Mercedes and BMWs and Audis that are tricked out beyond belief by people with a ton of money. I knew that the sight of a rat rod among those supercars would draw all sorts of crazy attention. So I gave Aaron his task: "Dude, it's got to have the look, the rust, the patina, but I want to be able to go 150 to 160 miles per hour. I want A/C, I want big brakes—I want this to be the s—t!"

Aaron dove in and he basically built that car himself. I mean, he was the only guy that we had, really, other than a couple of part-time helpers here and there. And when I ran it in the Bullrun, it got more press than I even imagined. I was doing a buck seventy down the freeway in a car that was one inch off the ground with no paint on it. I just knew that every other guy in the business wanted to get himself some of that!

I was my own promoter for all that time, too. I was always making calls to the magazines and TV shows and anybody who'd listen. I'm a big proponent of *If you think it, you believe it, and you get it out there, it'll come to you.*

I was also the one making phone calls to production companies and TV networks, trying to get somebody, anybody, to come out and take a look at what we were doing. A company called Pilgrim Studios was the force behind *American Chopper.* They also produced *Ghost Hunters* and *Dirty Jobs with Mike Rowe.* As far as I was concerned, they were the only company to go to with my TV show ideas, and so I called them up. Every week or two. For *years.* I was absolutely relentless. Nine times out of ten nobody would pick up the phone, but then every once in a while, somebody would get on the phone with me and share some tidbit of information that would get me one step closer. The main thing I needed, as far as I could tell, was to keep developing my street cred while putting together a "sizzle reel"—a short, well-produced video that would show off Gas Monkey Garage and what a potential TV show would be all about. I wasn't sure how one might go

The Gas Monkey Bullrun rat rod. *COURTESY OF RICHARD RAWLINGS.*

about getting a "sizzle reel" made, but I was confident I'd figure it out eventually.

Overall, my pushing seemed to be working. Unfortunately, though, all of my salesmanship, grandstanding, rule-breaking, attention-grabbing antics caused a lot of pushback among the "purists" of the car world. I put "purists" in quotations because there's no such thing. A lot of the pushback I got was complete BS—but such is the nature of the game. Especially since the Internet was newly booming with bloggers, and for some reason people started giving bloggers a whole lot of power and sway despite the fact that most of them had no credentials or worth whatsoever. It was a battle I never expected. It caught me off guard. I tried to ignore it, but I also paid a price for not being better prepared for it.

The message boards were pretty rough on me in those first few years. There'd be all kinds of comments from people after a show or a rally saying, "Oh, that Richard Rawlings guy sucked! He's just some rich dude with a trust fund and blah, blah, blah." One of the biggest message boards in the hot-rod world is called the HAMB board, which I refer to as the Hokey Ass Message Board. It's found on the Jalopy Journal, and it's considered one of the more respected sites. Back then, it was very respected because it was a couple of guys out of Austin who were really into hot rods, and they had built this thing that nobody really knew. It was kind of the first big, giant blog thing that all these car guys were on. And all these car guys were negative just because I made such a big splash so quickly, you know? "Oh, f—king Gas Monkey and the trust fund strike again." I mean, there were all these false stories out there about me being a bazillionaire with my own helicopters and islands and everything. It was nuts!

There was one particular guy who was always instigating problems and I didn't know how to handle it at first. I didn't have a public-relations consultant working for me. All I had was me! So I responded one time, in earnest: "Hey! You're talking about my company and my money. This is all me. I don't have rich parents. I don't have anything."

He turned that back against me and then actively started getting all sorts of other guys riled up about bashing me on those boards. Next thing I knew it was ruining my marketing plan and affecting my business. Companies and individuals didn't know what to make of the blogs at first. They didn't realize that most of them were just one whiny loser in a basement somewhere making noise in order to get attention. They believed what they read!

I learned pretty quickly that responding to haters online only gets you more hate. So instead I responded in person. I found out who this guy was, and I tracked him down at one of the big hot-rod shows one time. He was sitting there selling hood ornaments and vintage suitcases or something, like some pauper on a corner.

"You and me got a problem?" I asked.

He was shaking.

"No, no problem."

You see, I'd done some research on the guy, and I knew that one way he made some of his income was by working as a photographer, shooting photos at all of the big car shows. The thing was, nobody knew that this photographer they were hiring was actually this troll who caused everybody problems on the Internet. So I called him on it, and I basically threatened to expose him for who he was. Without any fists flying or anything else, he quit bashing me.

I ignored the boards as best I could—until another round of bashing cost me a television deal. I studied up and did some investigating and found out the guy who was in on it was literally some kid who lived in his parents' basement. No joke.

Aaron and I built a rat rod for Corky Coker, of Coker Tire, and it made the press. Discovery was interested in turning that build into a special, too—until the blogs messed it all up. They started bashing Coker. And then they got wind of the Discovery interest, and they started bashing Discovery, too!

There were cease-and-desist letters sent from lawyers on all sides, and none of it led to any resolution. It nearly crushed me. I couldn't believe that these sites could have such a dramatic effect on my business, and the business of these big, powerful companies. People really believed what they read on the blogs in those days, and it set me back more than I ever could have imagined. There was nothing I could do. I thought about dropping everything and just beating the living daylights out of those guys. But in the end, I knew it wasn't worth it. I just had to suck it up, stay back, and hope that they'd eventually move on to a different target. But to this day, those terrible, mean-spirited, completely untrue things are still out there.

I knew then and there that the bigger my business became, the more I'd need to get out in front of everybody else on the Internet. Of course, I never could have imagined that blogging would lose all of its steam to the immediacy of social media. But when those changes came along, I was ready. Going through that difficult period prepared me for the future that I hoped was still on its way.

I did my best to move on and not focus on it. I went out to more shows. I kept doing rallies. I was right in the middle of another Bullrun in 2007 when a buddy bet Dennis and me that we couldn't beat the world record for the Cannonball Run, and I abandoned that race from up in Canada to head straight to New York City and make the driving run of a lifetime.

I was flying high when Dennis and I accomplished our goal.

Thirty-one hours and fifty-nine minutes! It was a rush, man. Dennis and I both were so pumped that we accepted yet another bet to go run another rally down from Miami to Key West immediately on the heels of that. That race got crazy and involved helicopters and planes and all kinds of crazy tricks—I could write a whole book about that week—but we'd win that one, too! Getting on *The Tonight Show with Jay Leno* was the capper to all of it. I was getting all of the hype and attention on a national level that I'd been aiming for from the start.

Still, that negativity online hurt. I was deflated by it. When Discovery backed out of that special, I thought about shutting the whole business down. I really did.

And then the economy tanked on top of everything else. Business at Gas Monkey Garage leveled off. I wound up selling the big rig just to help with our monthly budget. It had done its job. It had pretty much run its course. I was sad to see it go, for sure. But I wasn't about to give up. I kept going to shows. We kept flipping cars, using all sorts of advice from Dennis and my own personal experience to keep turning a profit even at the worst of times. Still, the only word I can come up with to describe my mood during that late-2000s period is *deflated*. The air seemed to be draining right out of my Gas Monkey balloon.

Life sure can be a roller coaster, can't it? Just as I was hitting a real low and wondering if Gas Monkey was ever going to make it, I wound up talking to a guy at a car show who was launching a new clothing company. I offered to wear all of his clothes on the next Gumball, but there wasn't enough time for him to get them to me. They were only in prototype mode, making clothing from fabric from old car interiors. That conversation took an interesting turn really quickly, though, when I told him about my ambitions for turning Gas Monkey Garage into a TV show on Discovery. (Just like Pilgrim was the studio I wanted the most, Discovery was always the network I wanted the most. Always!)

"I think I can help you with that," the guy said.

"What do you mean?"

"Well, we've got a little production company up here, and we're trying to break into bigger things. So we could shoot your sizzle reel for you," he said. "After hearing what you're up to, I'm gonna say we can do it for nothing," this guy said. "It's worth the gamble for me."

Seriously! That opportunity just dropped in my lap—all because I didn't give up, and because I wasn't afraid to talk about my big ideas, even when things didn't feel so big at that moment.

Right after I came back from Gumball that year, we shot a big sizzle with me and Aaron. The show concept at that time was more about what we were doing, which was finding cars, changing them enough to make a profit on them, and reselling them quickly. It was also about hot-rod culture, as we took the cameras to hot-rod shows and talked to all kinds of interesting people at the shows.

I got that sizzle to Pilgrim—and Pilgrim liked it. After all that time, they put an option on it to shop it to cable networks! Finally things were looking up.

Of course, sometimes when things look up, that only means you're about to hit the crest of the hill. That sizzle didn't land us a show. It kept hitting wall after wall after wall. The roller-coaster ride continued.

It was during this same period when I let the Gas Monkey notoriety go to my head a little bit too much. I was so focused on work—and on the partying lifestyle that came along with life on

I wanted to be Burt Reynolds in *Smokey and the Bandit* even way back then! I'm pretty sure that cowboy hat was a Christmas present. *COURTESY OF RICHARD RAWLINGS.*

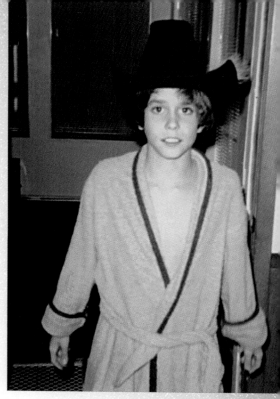

That's me with my sister, Daphne, who's modeling a dress from Funky Designs. At one point that company provided dresses for *Dance Fever* with Deney Terrio, which means we were like two degrees of separation from TV fame even as kids! *COURTESY OF RICHARD RAWLINGS.*

I think my dad was shocked that I actually graduated high school. *COURTESY OF RICHARD RAWLINGS.*

Some guys peak in high school. Clearly, I was not that guy. Check out those Zodiacs on my feet, though. Those were pretty stylin' back in the '80s! *COURTESY OF RICHARD RAWLINGS.*

That's my wife, Sue, on the back of my '97 Harley Davidson Softail chopper, during one of our many road trips. *COURTESY OF RICHARD RAWLINGS.*

That's me in the middle with my buddies Dennis Collins (left) and Jay Riecke, exhausted after a rally. *COURTESY OF RICHARD RAWLINGS.*

That's the old Gas Monkey rig in all her glory, parked in front of the famous Mann's Chinese Theatre in Hollywood. Aaron and I practically lived in that thing during the first few years we spent building GMG's reputation. *COURTESY OF RICHARD RAWLINGS.*

A peek inside during Gas Monkey Garage's earlier days. *COURTESY OF RICHARD RAWLINGS.*

Here's me and Aaron on our first full day of filming for *Fast N' Loud,* looking to turn rust into gold in a field full of old cars and trucks. *COURTESY OF DISCOVERY COMMUNICATIONS.*

I managed to get that new sign designed and hung on the side of our tiny rented shop with just five days notice before the cameras arrived and *Fast N' Loud* started shooting. *COURTESY OF DISCOVERY COMMUNICATIONS.*

That's the original Gas Monkey logo, on the back of the very first car Aaron and I ever built. I was able to buy it back recently, and I keep it in the garage now just outside of my office. *PHOTO BY MARK DAGOSTINO.*

This here's another early Gas Monkey build that I was able to buy back for my personal collection—a rat-rod style Model A. *PHOTO BY MARK DAGOSTINO.*

That's me and Aaron with the '71 Kingswood station wagon that we turned into a dragster. We sold that car to Ugly Kid Clothing, and it was right in line with the anti-bullying stance of that whole brand: we took that ugly duckling of a car and showed the world just how cool it could be! *COURTESY OF DISCOVERY COMMUNICATIONS.*

Aaron and I ran the Bullrun in this Chevy Caprice we built. We called it the Crap-Piece 'cause it never finished a race. Ever. We had more tow bills than build bills on that car. *COURTESY OF RICHARD RAWLINGS.*

My Harley, right where I park it in the garage these days, next to one crazy-looking chopper I picked up for myself on eBay one night. *PHOTO BY MARK DAGOSTINO.*

One of Aaron's proudest moments—taking his Falcon to the top of Pike's Peak. *COURTESY OF DISCOVERY COMMUNICATIONS.*

I keep this $200 Pinto just to remind myself that *any* car can get turned into a hot rod. How many Pintos do you see on the road today? They're so uncool, they're cool! *COURTESY OF DISCOVERY COMMUNICATIONS.*

Blowin' off steam while burnin' some rubber in the GMG parking lot. *COURTESY OF DISCOVERY COMMUNICATIONS.*

A before-and-after of the VW shorty. Fans loved this chopped and squeezed little van. Truth is, it may be fun to look at, but it's not really safe to drive over 25 MPH. The thing could tip right over! *COURTESY OF DISCOVERY COMMUNICATIONS.*

Aaron built this bike for a tire shop back before we were on TV. I later bought it back just to keep it in my personal collection. He fabricated just about every part of it from a vision in his head. No drawings, no computers, nothing. That's pure talent right there.

PHOTO BY MARK DAGOSTINO.

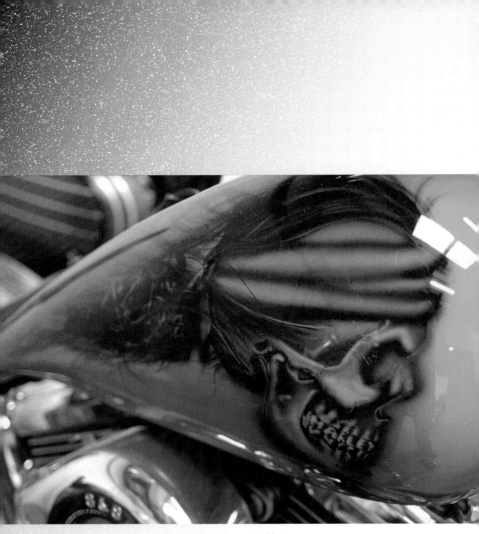

A close-up look at the custom paint job on my Harley. *PHOTO BY MARK DAGOSTINO.*

I keep this framed collection of photos on the wall right next to my desk, featuring multiple views of my gold '65 Mustang 2+2 fastback—the car I got shot in! *COURTESY OF RICHARD RAWLINGS.*

Here I am up on stage with a Miller Lite in my hand, talking to a packed house full of fans at the Gas Monkey Bar N' Grill, right here in my home state. I tell you, folks: Life don't get much better. *Whooo! COURTESY OF RICHARD RAWLINGS.*

the road—that I neglected my marriage in the worst way possible. Sue and I got into some heated . . . let's call them "debates" over what I was doing and how I was spending my energy. Finally she said she wanted a divorce, and I moved out that day. I left all of my stuff behind and bought brand-new everything. I wanted a clean start, and I certainly didn't want to stay with anyone who didn't want to stay with me. I loved her, but I screwed up, and I accepted the error of my ways. There was no fighting from that point forward, as far as I was concerned. We'd always kept our businesses and finances separate, and so we just split everything the way it was and went our separate ways.

It wouldn't last long. We'd start spending time together almost immediately after the divorce was finalized. She really is an incredible woman, and I will question why I did what I did probably for the rest of my life. We tried to be friends, and sometimes we'd be more. But I had a feeling we'd always be in each other's lives no matter what the status of our relationship. That feeling would prove to be right in ways we never could have imagined over the next couple of years.

On the business side of things, after we got that sizzle reel optioned by Pilgrim, two more years went by. We kept Gas Monkey alive primarily by flipping cars. Most of the money that came in went right back out. I wasn't pulling a salary for myself, and most months it took everything I had just to keep paying Aaron.

Pilgrim stayed in touch and seemed enthusiastic about the show's possibility, but they finally called and said they wanted a new sizzle with a slightly different concept. So we went out and put it together, and the show morphed a little bit more into the car-build show you see today instead of the road-tripping, presto-change-o car show we thought it was going to be. That was a good thing. It forced us to go out and drum up more business that would put Aaron's true talents to work. We were super happy with the way the sizzle turned out, so we sent it to Pilgrim, and then we waited again. I kept calling them every week looking for updates, looking for some bit of good news to hold on to. And week after week I was told that our show was "still alive!" but that there wasn't any news to report as of yet.

Part of me wondered if that was just Hollywood's nice way of telling me to get lost without actually saying it. Hollywood's like that. They talk out of both sides of their mouths and try not to burn any bridges, just in case anyone ever becomes successful. They want to be able to come back and be all supportive and ride your coattails if and when it happens. It's just not as honest a place as Texas, that's for sure. I'm not disparaging anyone in particular. That's just the way Hollywood operates.

Finally, sometime in mid-2010, I decided I'd had enough. The shop was barely holding on, and it just didn't feel like very much fun to me anymore.

It wasn't that I was quitting. I would never, ever quit. I just needed a bit of a breather—and a chance to make some cash to maybe get this thing fired up again.

Aaron Kaufman, with all of his talents, wound up taking a side job at a regular everyday four-wheeler shop, installing shocks and fog lights on pickup trucks. It killed me to see his skills going to waste like that. So we moved the operation into a smaller garage, one that had an apartment above it where Aaron could live and work and take on custom jobs and the sort of work he loved until I (hopefully) got Gas Monkey back on track.

In the meantime, I went to work for my ex-wife's home-health-care company and I tried to make the best of it. Like I said, Sue and I would always be in each other's lives. I just didn't realize we'd be working together so soon after we split, or that my own

Aaron, alone at Gas Monkey Garage . . . *COURTESY OF DISCOVERY COMMUNICATIONS.*

business would come so close to going bust. I was living in a lousy apartment while I threw my business acumen into Sue's company as if I were building a new company on my own. In some ways, doing that helped me to refocus myself. I helped her company become far more profitable than they'd been up until that point. It was good work. I liked that I was helping Sue become more successful, and I pulled in some good income for myself in the meantime.

At some point a few months into this, Pilgrim Studios called once again and said, "Hey, would you do one more sizzle? I think we got some traction with Discovery." So I called Aaron and we did a third sizzle reel, which focused even more on the car-building aspect, getting the cars sold, and doing it on a tight timeline to meet the needs of some incredibly demanding customers. Neither me nor Aaron thought anything would come of it. We just did it for the hell of it. It was kinda fun. We'd given it a good shot. We'd had a blast while we tried.

Putting an end to all of the promotional work was certainly easier. Aaron is definitely more comfortable just doing his thing in the garage. I was working nine A.M. to three P.M. most days, and I very quickly got back to pulling in a six-figure income. I wasn't on the road anymore. I actually saw more of my ex than I'd seen of her in the last couple years of our marriage.

I hate to admit this, but I was starting to feel comfortable with the thing I dread most in life: *settling*. When we didn't hear back from Pilgrim for months again, I was almost, *almost* ready to give up my Gas Monkey dream and just say, "F—k it."

And that's when the phone rang.

PEDAL TO THE METAL!

At the old garage, at the start of *Fast N' Loud*! COURTESY OF DISCOVERY COMMUNICATIONS.

I picked up my phone and it was Craig Piligian, the founder of Pilgrim Studios.

"Discovery wants the show. They're buying six episodes," he told me. "They need to slam this into the schedule, which means we need to start shooting the pilot five days from today or the deal's off."

"Craig," I said. "There's nothing to shoot."

At that moment we didn't have any cars in the pipeline that were anywhere near the super-cool interesting stuff Aaron had been building a year or two earlier. We'd really, really dialed it back!

"Well, it needs to be up and running in five days. Do whatever you need to do."

"In five days?"

"Five days."

I didn't know whether to get excited or to give up. So much time had gone by that a part of me wondered if I should just tell him, "Thanks, but no thanks. I'm not interested anymore." The five-day time frame didn't worry me. I knew I was capable of doing anything under any deadline if it came right down to it. I just wasn't so sure that I wanted to pursue that dream anymore. I was about as defeated in that arena as a guy could get. I'd already dedicated nearly eight years of my life to chasing a dream that kept running away from me again and again and again. I was pretty invested in building my ex's business now, and I was making good money. I honestly asked myself, *Do I really want this anymore?*

I told him I'd need to think about it, which didn't make him very happy. But then I went off and discussed it with Sue. That's right, I went to my ex-wife to talk about one of the biggest and potentially life-altering decisions of my life. It's a pretty strange thing to do for most people, but I knew that she was the only one who fully understood what I'd invested in this, and just how much of a toll the whole endeavor had taken on me.

"We could bend over backward to do all of this, and all they're giving us is six episodes," I said. "We don't have any control over the edit. I mean, it's just a full-on shot in the dark."

Was it worth putting Aaron through that? Was it worth putting myself through all of that for something that was a longshot? At that point, if the show didn't take off there was a chance Gas Monkey Garage would just close up for good.

That's when my ex said the one thing I needed to hear: after all of the time and energy I'd spent trying to build the Gas Monkey brand and business, she said, "If you don't do this now, you'll hate yourself for it."

There's a reason I married that woman.

She was right.

"You know what?" I said. "F—k it. You're right. It's a *chance*."

It was more than a chance. At the very least, I'd be able to end my Gas Monkey odyssey by saying I actually got a show on the air. How many people can say that? I *had* to do this. I owed it to myself. I owed it to Aaron. I owed it to the small but powerful Gas Monkey fan base we'd developed around the country in our years on the road—all those people who wore our T-shirts and who came out to support me at the Gumballs and Bullruns. I needed to do it for them. I needed to do it to prove all the haters wrong. Most of all, I needed to do it because it was my dream.

All of a sudden I got angry at myself for almost giving up! What the hell was I thinking? I'm not a guy who settles. I'm no quitter. I picked up my phone and called Piligian. "Hell yes!" I said. "I'm in. Get your crew to Dallas and let's roll. *Whoooo!*"

You know what's kind of amazing? Sometimes things work out in life, and sometimes they work out even better than you plan.

My old shop was big and cushy and comfortable and could hold all of the equipment Aaron needed to get the job done on just about any vehicle. The new smaller spot we moved into? It was way too small, a little bit dingy, and had nowhere near enough room for everything we needed. We were starting over from scratch and just scraping by. Capturing the drama of how we were going to get this business up and running again with minimal resources made the show more interesting! I could see that from day one. And I could see that it would give us room to grow as the episodes went on. With a little time, we'd be able to move into a bigger and better shop, and that would be fun for the viewers to watch—assuming we made it past the first episodes, of course.

Once we got this thing rolling, though, I got right back to being the big-dreaming, goal-oriented Richard Rawlings I'd been all along. With a dozen or so crew members around pointing cameras in our faces while we went about our business, I knew where I wanted this show to go, and I knew that I wanted the audience to get invested in growing the business right along with us. It all made sense. We were starting over—not from scratch, but from what we thought might be the absolute end of our business. We really would be building it back up from here. Every job we did we really needed to make money. We'd need to flip some cars for quick cash just to pay for expenses. We'd need to be smart and fast, and all the more dedicated to working at breakneck speed, because the budget would dictate our timelines now as much as, or maybe even more than, any customer we could drum up. We'd need to produce a full hour of television, complete with A stories and B stories (that's what we call the main story lines and the secondary, smaller story lines), every two weeks. There was no wiggle room. If we didn't get shows in the can on time, we risked losing our deal. Period.

I suddenly felt the enormous pressure of knowing that I'd really only have a few episodes to get this right. If the network or the viewers hated it, there would be nothing I could do about it.

As you can imagine, the five days leading up to the first day of shooting were some of the most intense of my life. I suddenly had to refocus everything I had on what would work—and what wouldn't work. I didn't want our show to be filled with all the infighting and nastiness that you'd see on so many other reality shows. I wanted us to focus on the personalities, but also on the cars themselves. Cars have great history. Cars bring up all kinds of nostalgia for people. They're an important part of our childhoods. They're an integral part of our daily lives, and a lot of people dream about tooling around in a convertible Mustang or some other cool car at some point in their lives. It's a common dream. We needed to tap into that.

With just five days to think about it all, I realized that I needed to make a couple changes. One: I needed to cut my hair. My hair was so long, I'd been wearing it in a ponytail. For years. But I knew that ponytails were perceived as some sort of hippie-rebel type of thing, and that wasn't really the image I wanted people to see on TV. It was a part of who I am, of course, but it wasn't *all* of who I am. How you look makes a powerful impression on an audience.

The guy who cuts my hair had been encouraging me to snip

it off anyway. When he heard that my TV show was finally a go, he insisted it was time. I'd been wearing that ponytail for more than a decade. It was a huge thing for me to cut it off, and it took me a bunch of hours and a whole lot of Miller Lite before I finally relented and said, "Do it!"

Snip.

That was it. One swift clip and it was gone. *Boom!* A whole new me.

After making such a dramatic change, I decided to leave the rest of me exactly as I was—tattoos, rings, ripped jeans, and all. I needed to be aware of what the audience wanted, the same way somebody might dress extra nice or do their hair a certain way before they go into a job interview. But I also knew that I needed to be me. Authenticity comes across, and I never wanted to seem like anybody other than who I was.

The even bigger issue I suddenly felt the need to change in those five days, though, was the Gas Monkey logo. We'd been running around with the same basic logo for almost ten years, and it suddenly occurred to me that it just didn't capture the brand. It was a great car-shop logo. It was rebellious and edgy looking. But I needed to have all of that combined with something much more family friendly.

I called a designer pal and a couple of friends over and we sat around a big whiteboard and sketched out some ideas. I liked the Gas Monkey name, and those guys got me talking about what it was all about. In addition to being about guys (and gals) who enjoyed cars more than almost anything else in life, I really liked what the monkey represented. A monkey's a bit of a rascal. He's curious. He likes to get in there and monkey around with things. He's into having fun just for the sake of having fun, too. He's a troublemaker. But he's also so darned cute that kids find him all sweet and adorable no matter how big of a rascal he may be.

That was it. The monkey needed to be the whole focus of our logo. The skull and crossbones, swords, any kind of symbols that might elicit something from wartime (the way West Coast Choppers did, which caused a lot of controversy a few years down the line) had to go. My designer sketched out a crazy-looking monkey with his tongue sticking out, and I knew it was perfect. It was just a gut reaction, with no time to ask for anyone else's opinion other than my own. I ran out and got some big signs made to tack onto the side of the garage. I made new business cards and some bumper stickers, too. And then I rush printed some T-shirts for my employees to wear on air.

Oh yeah! That was the other thing. We needed a crew! On super-short notice, we hired K.C. to be our paint guy, and Scot McMillan to come in and work as a fabricator to help get this show on the road.

K.C. was an old high school friend of Aaron's, and Scot knew Aaron from back then, too. I took his word on it, met these guys quickly, told them what we were doing, and made them an offer to join the crew—knowing that we were only guaranteed six episodes. It basically meant they could be out of a job in a few weeks, or, if all went the way I planned, they could be setting themselves up to make more money than they could possibly make at their current jobs, and gather some fame and notoriety to boot. It was a gamble for sure. A big gamble.

In K.C.'s case, he had a good job as a service technician for Coca-Cola when we approached him. He had a wife and two kids at home, and he had to make a decision about joining us in something like a twenty-four-hour time frame. It was crazy to even ask a young man to think about making that kind of a choice. Lucky for us, his wife had a great attitude about it. She told him, "Well, if you wind up out of a job after three months and you can't get your old job back, you'll find another job. But this sounds like the opportunity of a lifetime. If you pass it up, you'll always regret it."

K.C. took his wife's advice, and Scot said yes, too, and suddenly we had a crew! I knew we'd need to make some more hires, but we'd have to do them on the fly as the show started taping, because we were down to our last twelve hours before the cameras would be at the garage and the first day's shoot would be under way.

Next thing, I went out and hired the best social-media experts I could find. I knew I needed to tap into Facebook and Twitter big-time. I needed to get ahead of any bloggers that might pop up to say disparaging things about me. I needed to make sure the whole world could find Gas Monkey Garage on Google. All of it. And I needed it done fast! I was all pumped up about it as I brought this fantastic girl named Lauren into my Gas Monkey circle. I told her I wanted a million followers as fast as humanly possible, and she talked me down and explained how the number of followers Gas Monkey had wouldn't be nearly as important as how engaged we are with each follower we have. I guess that one strength I have as an entrepreneur is that I know when to listen to those around me who have more expertise in an area than I do. I listened to Lauren's advice, and I got a crash course in how to send my own tweets and take better photos with my iPhone so I could be ready to interact with Gas Monkey fans. Luckily, we already had a website up and running where we'd been selling Gas Monkey T-shirts for years. But Lauren and her team helped us get the whole thing revamped and up to speed as quickly as possible.

Finally, with about twelve hours to go before we started shooting the first show, I sat down with Aaron and said, "Well, what should we build?"

Of all the cars we'd built and fixed and flipped at Gas Monkey since we started, we knew that the Model A—the classic hot rod of all hot rods—was the one we'd done most often. There's an estimate I'd once heard that there was one Model A out there for every 1.1 square miles of earth in the United States. That's how prevalent those cars were back when Henry Ford first ruled the automotive world. They were cheap. They were common. They were easy to fix up and modify. And because they were cheap back when people first bought them, people often abandoned them in a field or left them in a barn somewhere when the engine conked out. There was no sense in fixing them. It was just as easy to go buy a new car. That meant there were Model As all over this country just waiting to be found and brought back to life—or "reincarnated," as Aaron likes to say.

I hopped on eBay, and much to my surprise, I couldn't find one. I was shocked. I started to panic. *What are we gonna do if I can't find one?* We didn't have an alternate plan. I'd kept about ten hot rods in storage, so maybe we could come up with something to do with one of those, I thought. But how boring would that be? The whole idea here was to show people that amazing moment of finding a diamond in the rough, a gem of an automobile that was hiding in some remote location just waiting to be uncovered. I loved that feeling. I needed to share that with the audience!

It was early the morning of our first day's shoot when I finally found a Model A for sale that looked like a perfect car for us to work on. Sure enough, it was in a barn up in Missouri, if my memory serves me right. That meant Aaron and I could hop in my truck and go get it together. A road trip. An adventure with a prize at the end of the journey.

The first-ever episode of the show that would be titled *Fast N' Loud* was under way. It was time to let go, to have some fun, and to dedicate myself to cars full-time. I was finally completely embracing the passion that had bitten me way back in my teens.

And guess what? Discovery loved it. Before we finished building our third car for the series, the network ordered six additional episodes. By the time episode number three aired and the ratings came in, they ordered nine more.

All those years of hard work and struggle finally paid off, thanks to you, the Gas Monkey fans who said loud and clear, "Hey! We love what you're doing! Keep it up! And give us more!"

We've been shooting episodes one after the other for nearly three years straight since that first episode got under way, and all told, we've only taken about two weeks off in all that time. It's crazy! I love it. I couldn't ask for a better job, a better career, a better lot in life. I'm living the dream, and so are the other guys and gals who make up the Gas Monkey family. Every one of them is enjoying the crazy trajectory we're on, as this show has become bigger than anyone (other than myself) ever dreamed possible.

Gas Monkey is much more than a garage now. It's a lifestyle. It's apparel. It's an expanding universe of awesome businesses that allow Gas Monkey's *Fast N' Loud* fans to get up close and personal with us. I couldn't be any prouder of what we've accomplished, and I couldn't be any more excited than I am right now to be sharing it all with all of you.

So let me take this opportunity to say, "Thank you."

You rock!

Thanks so much for coming along on this ride. Go grab yourself a cold one now and sit back for a ride of a different kind as I walk you through some of favorite episodes, show you around the ever-expanding Gas Monkey empire, and share a few tips and insights into how you can become even more of a Gas Monkey yourself than you already are.

As you've seen already, the key to making this all come true has been a journey of Blood, Sweat and Beers. Maybe that's the secret of life right there. What do you think? It sure makes one heck of a fun line on the back of a T-shirt if nothing else. But I really do believe it's that simple.

Whatever it is you're seeking to achieve in life, all you've really gotta do is picture it in your mind and then, "Get you some of *that*!" All you've gotta do is go do it. Dream it. Plan it. Resolve yourself to it. Put in the hard work. Give it your all. I'm living proof that anything is possible in life. Even landing your own TV show . . .

PART TWO
THE BIG SHOW

Aaron and me in the desert. *COURTESY OF DISCOVERY COMMUNICATIONS.*

ASSEMBLING THE CREW

Before we reminisce about some of our favorite episodes, I thought I'd say a few things about the core who's who here at Gas Monkey Garage.

Putting a crew together to pull off the work we do while simultaneously making good television is not as easy as it looks. All right, I'll admit, it must not look too easy. We've had a lot of turnover in the two years or so since the show started. K.C., Aaron and I are the only original Gas Monkeys who've been at the garage since the show began—primarily because a lot of people who think they can do this job can't handle the intensity of it. It's a lot of pressure. We shoot continuously, usually six days a week, with a week off in the summer and a few vacation days around Christmas and Thanksgiving. That's it. And our days aren't nine-to-five (as much as we try to make them nine-to-five). It often takes long stretches of work late into the night to get the builds done on time, and getting these cars out the door on whatever deadline we're working toward is the key to Gas Monkey's success.

From the beginning, Aaron and I determined we were going to build the best cars faster than anybody else has ever built them. And that's what we've been doing, week after week, with the help of our kick-ass crew. Sometimes we get wildly successful results, and sometimes not. I'm not afraid to show all of it—the good, the bad, the ugly, and the disastrously unprofitable. I insisted we needed to show people the real ups and downs of the car business. Nobody can win all the time. Nobody. And if we lose money on a car, we're gonna tell you.

Anyway, the magic formula I've uncovered for putting together a solid crew—whether it's at Gas Monkey or a print shop or what have you—is that there *is* no formula. It's all trial and error. With the enormous pressure we're under, it's important that everybody in the garage gets along. Sometimes you just can't figure that out until you get people in the door and see how they perform when the heat's on. Or when the heat's off, as it were.

That first garage space we were in didn't have any heat—and we started shooting in the middle of winter. It got cold! It didn't have any running water, either. We had to run up to the building in front of ours to use the bathroom or wash our hands. I had to rent a little office trailer to park in the parking lot just so I could have a spot with a computer and a phone in it. It was nuts. And for the whole first three months of shooting, Aaron, K.C., and Scot were pulling twenty-two-hour days just trying to get the work done on time. No joke, twenty-two hours a day. For the six months after that they pulled like twenty-hour days, and that felt like a frickin' vacation compared to what we had been doing.

The guys often slept in the paint closet, which was the warmest spot in the shop. We put a couple of cots in there, but they had to sleep in cold-weather sleeping bags to make it through the night. I remember coming in at six o'clock in the morning and

A hard day's work at
the original garage . . .
*COURTESY OF DISCOVERY
COMMUNICATIONS.*

. . . and what it looks like on an average day
inside our new spacious digs!
PHOTO BY MARK DAGOSTINO.

seeing K.C. wrapped up like a caterpillar in a cocoon, with the
hood around his head and the sleeping bag zipped up so all you
could see was his nose and his eyes. Sometimes they'd take naps
in K.C.'s truck, too, working in shifts just to keep their momen-
tum up. At one point K.C. told me his truck kept running in the
parking lot for twenty hours straight, just so they could hop in
and get some heat for a half hour as the day dragged on.

You put a whining, complaining, half-assed sort of worker

into a situation like that, things are gonna get ugly. I definitely felt fortunate that those guys got along during the difficult start to getting *Fast N' Loud* off the ground.

The fact is, everybody who stays at this shop is cool. There are times when we get angry or frustrated, and there are certainly some shouting matches. But even in the midst of all the hard work we put into getting cars built and putting a TV show together, I always make sure the crew has a lot of fun, too. The whole staff tends to go grab lunch together five days a week. It's not uncommon for me to call for a "F—k-it Friday!" where everyone on the whole staff just grabs a few beers or some shots of tequila after lunch and sits around shootin' the s—t. We get into some really stupid discussions on those Fridays, but you can hear people laughing from all the way out in the front parking lot by the Gas Monkey gift shop. It's awesome.

That sort of camaraderie just makes it all worth it. It really does. Whether we're pulling each other around on skateboards behind somebody's truck, or pushing all the cars out of the way so we can do donuts in the parking lot, the messing around and blowing off steam is a part of who we are. The little sayings that come out on this show—like "If we're gonna have fun, it better have a motor!"—that stuff's not contrived. It grew straight outta the fact that we're all having a good time doing something we love, whether it's chopping that frame back to get the air suspension just right, finding a pair of old Jaguars for sale on Craigslist, mixing the perfect color for a badass ride, or running the numbers and yelling at me to get my ass in line back in the office. Everyone I'm about to introduce here does his or her job really, really well, and has a whole lot of fun while they do it. That's part of what being a Gas Monkey is all about.

MEET THE MONKEYS!

Aaron. *COURTESY OF DISCOVERY COMMUNICATIONS.*

AARON KAUFMAN

Here's an undisputable fact: I wouldn't be where I am without Aaron, and Aaron wouldn't be here without me.

I hired Aaron to come into the shop and do what he does best because he's a talented motherf—ker, and I just can't do what he does. Nobody can do what he does. There are times I just stand back in awe, admiring how differently he sees things and how he puts things together.

I think the reason he has such a following among car enthusiasts is that he's self-taught, too. He just looks at cars from a different point of view than almost anybody else in the business. And he's very talented in figuring out a problem. He really is. He can look at something and figure out what's wrong. *Boom!* He doesn't need to open a book and read the instructions. Most of the time he'll buy something and open it up and talk about how it sucks and how he could've done it better. He's great at that, and that allows him to be more creative and do things to cars that most people wouldn't even think about.

A lot of his automotive genius doesn't even transfer to the show. You can see it in the results of how good a car looks, but you don't necessarily know what it is he did to it to make it look so cool. I mean, he can look at a car and figure out the geometry that'll make the wheels sit correctly, you know? When we're dropping something, he'll know how to get the car down closer to the ground than anybody else would even think possible, and he'll do it in a way that the car not only looks totally cool but rides great, too.

Aaron knows how good he is, too. He's very egotistical. He doesn't see the world except for how he sees it, and he's never going to take anybody's opinion or anything on it. Period. It's his way or no way. So that makes me have to walk very delicately

Aaron at work on a 1948 Fleetmaster. *COURTESY OF DISCOVERY COMMUNICATIONS.*

around trying to make a job come in on time, or make money.

The other thing that's remarkable about Aaron is he doesn't give up. He doesn't call anybody for help, either. He'll mess around with something for three days straight trying to get it right, which is fine except for he's taking up my time and money when he does that. I like to say that Aaron's been going to the University of Gas Monkey for eleven years. In all the time he's been working for me, on top of paying him, I've been giving him the opportunity to hone his skills. He was good before, believe me. But there's no other shop on earth that would have let this twenty-year-old kid just tear things apart and figure 'em out on his own. I saw how talented he was, and I wanted to see him keep growing. So before the show started filming, in that entire eight- or nine-year run-up of growing Gas Monkey, I made it a practice to say to him, "Just go figure it out and go do it." And most of the time, I didn't put any kind of time frame or constraint on him. I wanted and needed him to learn everything he could, so when the time came to build cars in record time, he'd be ready to tackle whatever got thrown our way.

Aaron's one hell of a quirky character, though, I'll tell you that. First of all, he couldn't care less about money. Everyone who works for Gas Monkey gets opportunities to go make money doing appearances and making endorsement deals nowadays. Aaron doesn't want to do hardly any of it! He truly just wants to spend his time building the best cars and trucks he can. Hell, if I let him, he'd live in his truck at the garage. We're talking about a guy who, for long periods of time during the eleven years he's worked for me, went without any permanent address. There are times when his electricity got shut off and I'd have to pay the fine for him to get it turned back on—not because he couldn't pay the bill, but because he didn't feel like walking to the mailbox to pick up his mail. "My mailbox is filled with all kinds of junk that gets sent to me that I don't want to look at!" he'll say.

His priorities are just different from most people's.

During times when he was sleeping in his truck or crashing at the shop, I would ask him: "Where are you gonna shower, Aaron?"

"Oh, I'll just shower at my girlfriend's house, or whatever," he'd answer. It just wasn't a priority for him to live like a normal human being.

Different. That's Aaron. The best car guy I know. The anti-establishment guy. Hell, I think the best time of his life was back when we were tooling around the country for months at a time in the Gas Monkey rig. Nowadays, I think he couldn't care less whether we were on a TV show or not. Truly. He just wants to build great cars. He wants his cars to speak for him, too. If he could avoid doing any and all interviews, I think he would. He wants his work to be his legacy and his crafts-manship to be his reputation. And I suppose that's what makes him so cool and appealing to the audience. Hell, it seems like every week there's some chick waiting for hours and hours out in front of the shop, just refusing to leave until Aaron pops his head out and waves at her or something. His fans are a dedi-cated bunch, I'll tell you that.

The last thing I'll say here about Aaron is this: Yes, the beard is real. No, he didn't have it back when we first met. You want to know why Aaron grew that beard? Let's just say he got in a little trouble with the law one time. As a sign of rebellion or something, he decided he wasn't going to shave until his proba-tion was finished. Of course, his probation got extended at one point and seemed to drag on forever, and then lo and behold we got ourselves on a TV show, and now everybody knows Aaron for his beard!

CHRISTIE BRIMBERRY

Aaron's fans may be dedicated, but let me tell you this: Christie's fans are passionate. Actually, the better word for some of Christie's fans might be *perverted*. We get dozens of requests every week to post pictures of Christie's feet on our Instagram feed. No joke! They ask for pictures of a lot of other things, too, which cracks us all the f—k up. I realize she's cute and all tatted up and she's got big boobs and all that, but come on, people! She's a mother of six! Tone it down a little!

One thing a lot of people don't know is that Christie is actually married to the guy who cuts my hair—the very same guy who's responsible for snipping off my ponytail just before we shot the first episode of the show. She was barely even an acquaintance to me before I hired her.

We were getting ready to shoot the second episode when I went in for a trim, and as I sat in the barber's chair I started complaining about how exhausting and overwhelming the whole process of the first episode had been. "I can't keep track of everything. I'm going nuts, man. I need an assistant like there's no tomorrow. So if you know anybody who's looking for work, please send 'em my way," I told him.

He laughed and said, "My wife needs a job."

He was just joking around, but I said, "Dude. Seriously. I'll talk to her. I need someone to start helping me out, like, now."

I swear to you, the moment Christie came walking in, I said, "You're hired."

She was like, "We haven't even talked yet!"

The fact that she gave me some attitude so quickly only reaffirmed my snap decision.

"Oh, yeah," I said. "You're hired."

Christie. *COURTESY OF RICHARD RAWLINGS.*

First of all, I only had to take one look at her to know that the TV cameras would love her—and I was even more right than I imagined. Hell, Hollywood should come see me about doing all the casting for their big shows from now on. Look what I did! She's a huge star now, with fans of all ages and both sexes. (Although, it's just men who write us asking to post pictures of Christie's feet.) But I could also tell just by the way she walked that this woman was tough. She was in charge. She had practice running a household full of six children, and that's pretty much what it's like to try to take care of me. Just ask my sister, Daphne.

Daphne was actually running the accounting over at my ex-wife's home-health-care company at the time or I would've just hired her to work as my assistant. I was pissed that I couldn't snag her away quick enough—until Christie came walking in. She wound up being a massively positive addition to the Gas Monkey crew.

First and foremost, Christie's a badass, and it takes a badass to be able to put up with me. She's also had to do her share of dealing with some of the crazy female fans who refuse to leave the premises in their quest to get to Aaron or me, and let me tell you: you do not want to mess with Christie. She'll cut a bitch!

She's smart, she's quick, she can multitask—she's truly the ultimate office manager and personal assistant anyone could ask for. She very quickly became not only my personal assistant, but my life coach and my confidante, too. The amount of laughing and joking we do around the office is almost criminal. You know what it is? My professional relationship with Christie reminds me a little bit of the one between Liz Lemon and Jack Donaghy, Tina Fey's and Alec Baldwin's characters on that TV comedy *30 Rock*. If you know that show, then I think you'll understand what I mean.

From the very first day, Christie jumped right in, tackled everything I threw her way, and then managed to give me s—t about how much most of the chicks I was seeing looked just like my ex-wife, Sue.

Every man should have a Christie in his life. Honestly. I don't know what I'd do without her.

K.C. MATHIEU

K.C. came into the mix and just fit in from day one. He also filled a very important role for us in the short time crunch we were

K.C. *COURTESY OF DISCOVERY COMMUNICATIONS.*

under to get Gas Monkey up and fully operational. We knew we were gonna need to be painting cars, and K.C. has a paint booth at his house. That allowed us to skirt all the rules and hoops we would've had to jump through if we wanted to try to paint cars at our newly rented garage space. It would've taken us six months to get the permits to do that. So K.C. was a lifesaver.

The fact that he's still here is a testament to what a cool guy he is. Honestly. He's the only one who's survived this ride from day one besides me and Aaron.

He's got a crazy fan base, too. Maybe they're drawn to the fact that he wears tall socks with his shorts or something. I dunno. But he gets recognized everywhere he goes. People are always shouting at him in his truck and asking him to do donuts. He'd be going through a new set of tires every two days if he answered every request to do a burnout in that truck of his.

The fact that he's been able to hang with us while juggling a wife and two young kids is pretty remarkable. The hours he put in during the first year alone would've caused most marriages to crumble. I guess his wife deserves a lot of credit for being so supportive of him taking this risk, too. Neither Aaron nor I were married when the show kicked off, and we both understand it's a whole different thing when you've got a full-time family at home.

K.C.'s dealt pretty well with some of the weird parts of this whole ride, too. Like the fact that fans sometimes look up his address and just show up at his house. Having some stranger knock on your door and act like they know you is a pretty weird experience. He's kept his cool, though, and seems to embrace the whole thing. He's also just really good at his job. When you think of the builds we've done, and how crucial the paint job is to our success, I'm just glad to have a guy like K.C. working on my team.

DAPHNE KAMINSKI

Daphne. *COURTESY OF RICHARD RAWLINGS.*

What can I say about Daphne? She's my sister. She's been with me since, well, since I was born, technically. She has worked for me at all of my companies, pretty much. Like I said, I couldn't take her with me right away when I came back over to Gas Monkey Garage because she was in a critical spot of the growth mode at my ex-wife's health-care company. But as soon as I was able to, I snatched her up and set her up here in the offices at the garage in Dallas.

There were originally no plans to show Daphne on TV at all. She was really just supposed to stay behind the scenes and take care of the books and business side of things. Once the cameramen caught a glimpse of the way she and I interact, though, it was just too tempting for them. We give each other s—t all the time and keep each other in check. It's just a fun dynamic for the audience to watch, because I'm sure everybody who's watching has a sibling like Daphne whom they get along with and bicker with in equal parts. The thing about having Daphne on board as a part of Gas Monkey is she's family. I know I can trust her. Always. She's great at keeping the books and doing all sorts of managerial tasks around the office that need doing. She also keeps tabs on my credit-card spending and reels me in when I'm spending too much money. (I have a tendency to do that sometimes.)

I sound like I'm repeating myself here, but truly: I don't know what I'd do without her. She's been there for me through all of the big ups and downs in both my business and my life. It's a pretty amazing thing to have her right there with me every day like that.

TONY TAYLOR

Tony very quickly became an integral part of Gas Monkey's success.

When I first met him he was running the service department at a local Lamborghini dealership, until it got sold. I met him through Dennis, and actually Dennis and I would sometimes go over and hang out with Tony and all of those $200,000 supercars on our Wacky Wednesdays, back when I owned the print shop. I gotta tell you, being around those cars makes you want one. That helped fuel my desire for big success right there.

Tony. *COURTESY OF RICHARD RAWLINGS.*

Anyway, Tony really takes care of a lot of the day-to-day operation. Now that we're a success, we get loads of calls and e-mails from people all over the country and all over the world who want to sell us their cars. It's way too much for me to handle by myself, so Tony's the guy who makes sure I see the really good offers. He steers me clear of the four-door Chevy Novas, or the 1972 rusted-out four-door trucks from Nova Scotia, and all of the crazy junk people are hoping we'll take off their hands. He filters through all of it to get to the good stuff.

Then, when we get cars in, he assesses them, does all the research, and double-checks the VIN numbers and everything to make sure everything's legit. "Is it really a '65 Buick 425 four-barrel?" or what have you. So Tony is instrumental in helping me filter through a lot of information that I just don't have time to sort through with the other craziness that comes with making a TV show and running Gas Monkey full-time.

Tony's also our go-to guy for searching out great cars, great deals, and great hard-to-find parts when we need 'em. The funny thing is, he didn't know much about doing any of that kind of stuff when he first started. He was a Lamborghini guy! But Tony's a quick learner, a hard worker, and he figured it all out and quickly became one of the best in the biz. That makes him a Gas Monkey through and through.

SUE MARTIN

All right, technically Yu-Lan Martin, aka Sue, isn't a part of our crew. She's just a lady up the road who does upholstery work, and who's always done a good job for us. She's also turned into a fan favorite. It's not uncommon for fans to stop by to see us at GMG and then drive up the road to see Sue at ASM Upholstery. Sue's great with the fans, too. It's kind of hilarious to think that her shop, which was right next to our old garage—close enough for us to push a car to if the engine wasn't running—has become a tourist destination.

One thing I've noticed though: Sue used to do all of Gas Monkey's upholstery jobs cheap and quick. Now, ever since *Fast N' Loud* started to get popular, her pricing is all over the place! She claims that I'm cheap. Well, you know what? Sometimes I am.

There's no question that Sue's interior work helped build the Gas Monkey reputation. So I guess that makes her an honorary Monkey. I could do without her calling us "Ass Monkeys" all the

Sue (with me and Aaron). *COURTESY OF DISCOVERY COMMUNICATIONS.*

time, but I'll continue to put up with it to remain on her good side. After all, they don't call her the "Dragon Lady" for nothing.

We've got a bunch of other great guys and gals working in the shop these days—from Mike and Jonathan and Keenan (who's been here since Season 1) back in the garage, to the girls out front selling T-shirts and merchandise to the public. I could sit here and say something about each of them one by one, but I gotta tell you: all this chatter is making me restless. I want to get back to the cars. I want to get to talking about some of the best episodes! So if you want to get to know more about the Gas Monkey crew, go check out our Gas Monkey Instagram and Twitter feeds, or go like our Facebook page. There's all sorts of fun facts and photos and videos and more on there from just about every one of us.

You won't find any pictures of Christie's feet, though.

Sorry, pervs. That s—t's just wrong . . .

OUR FAVORITE EPISODES

People always ask me what my favorite build was. You know what my answer is? My favorite build was the one that made me the most money! I try not to get too sentimental about much of anything when it comes to the vehicles we work on in the shop. If it makes me money, it's my new favorite vehicle. Every. Damned. Time.

I have become attached to certain builds, though, and I've even gone out and repurchased and retrieved a few that I'm really glad to have in my personal car collection. I'll talk about some of those a bit later. Of course, I wish I could get my hands on some of the cars I drove in my youth. I'd give anything to go reclaim my '65 Mustang fastback. I never should've sold it. I mean, how cool would it be to still be driving around in the car I got shot in?

Another thing people always want to know is what my favorite episode of the show's been so far. I've gotta say, my answer is pretty much the same. I don't have a favorite. I like 'em all! I'm more interested in which cars made me the most money. I'm working to try to keep this shop open and make a great TV show at the very same time, so I'd much rather leave the choosing of favorites up to our fans.

If there's a type of car we've worked on that gets me most excited, though, it probably always goes back to the '32 Fords. That's my favorite hot rod, period, because to me, that's where hot rodding started. When the guys came back from World War II, there were cheap old '32 Fords everywhere. They were easy to change over. They were easy to put new motors in and hop up and all that kind of stuff. We've only found and built maybe three of those '32 Fords in Gas Monkey history, so I was thrilled when we got to feature one on TV.

Besides that, it's hard to choose! So what I'm putting in this book aren't necessarily my personal favorites. It's really more of a look at *your* favorites. These are the shows that not only earned big

Checking out the '32 Ford three-window upon arrival at the shop.
COURTESY OF DISCOVERY COMMUNICATIONS.

ratings the first time they aired (thank you very much), but the shows that still draw thousands of crazy comments and tweets and Facebook posts every time Discovery decides to re-air them. What's crazy is how many fans out there will go online and argue about which episodes and builds are the best! I love it!

There are plenty of other episodes and specific builds and flips that aren't mentioned here, of course, and everybody's got a different favorite. So if you don't see one of your favorites here, please feel free to let us know. Head on over to our website or Twitter feed or Facebook page and make an argument for whatever you think was the best episode ever. We'll read 'em. We read every single word that gets posted. Plus, you can be sure to get in a heated argument with other fans about which episode's the best, too, because that's what people love to do on the Interwebs. Fight. Have at it, people. It's all good. The fact that so many of you feel passionate about which car or flip or show or individual Gas Monkey crew member is the best just means that we're definitely doing something right.

"MODEL A MADNESS"

I have to include this one, just because it's the first. Episode number one. The start of it all.

Man, I was nervous when this thing started up. I'd spent five crazy days getting everything in place just so we could start filming, and then we had to do this show when we didn't even know if we were gonna be on air, or if the show was gonna suck, or anything. It wasn't like we had this air of confidence as we started. We had more of an air of fear!

"Holy s—t! We're making a TV show! Make every second count! Try to make it good!" It was crazy.

The 1931 Model A, in a garage where it sat untouched for years.
COURTESY OF DISCOVERY COMMUNICATIONS.

Starting out with a Model A made so much sense for us. We do a lot of Model As. In fact, it's fair to say that Model As are what put Gas Monkey on the map. Heck, as I'm sitting here writing this, I had another Model A delivered just today. A '31 roadster. Indented firewall. All original, except for the wheels. Been sitting in a garage forever. There are still thousands—if not tens of thousands or hundreds of thousands—of Model As sitting in garages all over the world, but especially here in America. They sit in barns because they're not worth anything to people. They look back and go, "I bought that car for ten bucks." So they don't care that it's sitting there rotting away. I come in and buy it for $400 and they think I'm some sort of a sucker. They have no idea I'm gonna turn around and sell it to an interested buyer in some other part of the world for $5,000 or even $10,000.

Those really old cars always have a story to them, too. Even if you don't know the story, you know there's a story there and you can feel it. That's one of the reasons I love to chase them. There's all those questions, you know? How do they get here? Where do they come from? Why has it been sitting here for forty f—king years? What in the hell does somebody do for forty years that they can't go clean their car off and take it for a ride?

So there's that part of it. Then there's also the nostalgic part:

Admiring our finished product with K.C., Aaron, and Scot. *COURTESY OF DISCOVERY COMMUNICATIONS.*

Why did they design the car that way? Where did it come from? Why did the guys at GM or the guys at Ford or the guys at Dodge decide they were gonna lay the windshield back or they were gonna expand the doors or they were gonna make the car bigger, smaller, faster? There's so much there that is part of the history. It's one of my favorite things on the show, to tell them the history of where the car comes from.

Anyway, I found that Model A just hours before we started filming this first episode, and Aaron and the guys got started on a build that would happen faster than anybody else in the business could ever imagine. I knew I'd have a chance to sell it for a profit at the Good Guys Swap Meet in Fort Worth, which was happening just nine days from the day we got the car. So I thought that might make for some pretty good TV.

Apparently, I was right! Suddenly we were off and running. *Fast N' Loud* was on its way.

"HOLY GRAIL HOT ROD"

Six grueling months full of 120-hour weeks flew by as we cranked out the first eleven episodes of *Fast N' Loud*. Everyone was exhausted. We'd found and flipped and built more cars in that period than most shops would do in three years. We'd been through a Galaxie, a low-riding Lincoln, an amazing Impala, a '48 Chevy Fleetmaster, and more—and to be honest, I was finding it a little difficult to get fired up about what we might do next.

That's when I saw it. I did a double take when it popped up on Craigslist on my computer screen. I scrolled through every picture as fast as I could, I read the description, I immediately

The 1932 Ford three-window completed. *COURTESY OF DISCOVERY COMMUNICATIONS.*

e-mailed the seller and got his phone number. I simply could not believe it: a '32 Ford three-window. An all-steel car. The Holy Grail Hot Rod!

We'd hit a point of overtime with the film crew. They live by Hollywood union rules and such, and we were about to break early so everyone could rest up and start fresh the next day. I couldn't have cared less about any of that.

I ran out and grabbed Aaron and told the film crew, "We're leaving."

Everybody got in an uproar. "No, we need to schedule this, and da-da-da-da."

"This is a '32 three-window," I said. "So f—k y'all. I'm leaving!"

The Hollywood crew saw the error of their ways pretty quickly and jumped on board. They knew this was gonna be some great TV.

So off we went, making the two-or-three-hour drive up to Oklahoma City. I called the guy from the road. "Will you hold it for me until I get there?" I asked him.

"Well, you know, there are all these other people calling," he said.

"Look, I am on my way with cash. I will not beat you down on your price. If the car is what you say it is, I'm taking it!"

"Well . . . all right. I'll hold it for you," he told me. Still, I punched it. We hauled ass the whole way. I did not want to lose that car. And I didn't. We took that beauty back to the shop and Aaron put his personal spin on it. As it turned out, Aaron's personal spin wasn't what the marketplace was looking for at that moment.

For all the hard work we put in, and the incredible job Aaron did turning that car into a one-of-a-kind hot-rod that any one of us would've loved to buy for ourselves, I lost a lot of money when we took that car to auction. I poured $54,000 into that car thinking it would sell for upwards of eighty grand—and it only fetched $46,000! That made for good TV, I suppose, and we get tons of e-mails and tweets every time that episode re-airs, but losses of that magnitude were putting Gas Monkey in serious jeopardy. Not to mention it was causing all sorts of stress in the garage.

It was right after this episode that Scot decided to quit. He just couldn't take the pressure anymore, and I don't blame him. I really don't. Everybody had reached their breaking point.

THE ROAD TO *CHOPPER LIVE*

Aaron on our second-place-winning bike from *Chopper Live.*
COURTESY OF DISCOVERY COMMUNICATIONS.

The next two episodes would be dedicated to a special event we did, in which we'd build a motorcycle in direct competition with Jesse James and the Teutuls—the very builders whom I'd set out to best with my TV show from the beginning. It was a big deal, and a stressful time, and I'm not going to get into my feelings about how things went down or say anything negative about anybody. All I'll say is this: in the end we absolutely proved that we could hold our own and even beat those guys, despite the fact that we were still the new guys on the block. We beat out Jesse and Paul Senior . . . with a pink bike! Can you believe that Aaron built a pink chopper and still came in second place? That bike rocked. Aaron rode it all the way from Dallas to Vegas, too, just to prove that this wasn't some show bike for TV. It was a bike that any rider would be proud to own.

That whole event was a big, big victory for us. But, in the grand scheme of things, that build didn't bring any immediate income into Gas Monkey Garage, either.

We kept getting all kinds of great news about the ratings for the show. There were millions of people watching what we were doing! Still, the day-to-day reality of running the shop under the pressure of constant deadlines tested us all. The financial disasters seemed to keep coming one after another, and it really started to get to me.

Then? Things got even worse.

"MASHED-UP MUSTANG"

The Mustang on its way into the shop . . . *COURTESY OF DISCOVERY COMMUNICATIONS.*

I started out thinking this was going to be a great deal. Trading a '58 Impala that'd been sitting on my lot for too long for a '67 Mustang convertible with power steering, power brakes, and A/C seemed like money in the bank. We'd do some straight-up restoration, give it a little Gas Monkey attitude by painting it black and giving it a bit more of an aggressive stance, and sell it. Done.

. . . and after we'd given it a little Gas Monkey attitude. *COURTESY OF DISCOVERY COMMUNICATIONS.*

Then Jordan took that finished car out for a final test drive before we put it in front of our buyer, and some guy in a pickup truck ran a red light and smashed the hell out of our perfect moneymaker of a car. I lost my s—t. I was so angry. There were a few shots of me cussing out the driver that made it to TV, but the really choice words were said off camera.

Want to know what else happened off camera? Once the tow truck showed up, that red-light-running driver grabbed a bicycle out of his truck and pedaled away as fast as he could. He left the truck right there. He didn't want to risk getting into any kind of trouble with the law. He had no insurance. More than that, it's my personal suspicion that he may or may not have been in this country legally.

The Mustang was basically totaled.

What's really interesting is the way people responded once that episode aired. Our Facebook page blew up with people who absolutely fell in love with that black-on-black mustang—and who *cried* when it got wrecked!

My out-of-pocket loss on that car was $23,000. It's almost like the universe was trying to tell me something.

I decided to add a little postscript to that episode. When I was being interviewed on camera, I mentioned how that Mustang incident opened my eyes. I said, "Things are changing at Gas Monkey, and changing now. It's time to step this game up."

I knew I needed to turn things around, and I came to a big conclusion: we'd been playing it too safe. We shouldn't be messing around doing traditional restoration projects like that. We shouldn't be going for the easy money on builds. That wasn't the Gas Monkey mission. We needed to go bigger and bolder in everything we did. I wasn't about to sink because of these challenges, either. That's not the Richard Rawlings way, and it's certainly not the Gas Monkey way. Instead, I needed to take these challenges head-on and learn a valuable lesson from them. What we needed to do was to take on bigger risks, more-demanding clients, and put everything on the line—just like I'd done in the beginning.

What you didn't see on the show was the after-hours moment when I gathered the whole crew in a big circle and thanked them for the hard work they'd done. Then, I told them: "We're about to work a whole lot harder."

That elicited a few groans, of course.

"But," I said, "we're also gonna work a whole lot smarter."

By the way, for those of you who still cry over that car, here's some good news: After further assessment and a lot of painstaking work, we were able to get that Mustang back together again after all. It would never be the same, and all of those additional man-hours would still mean I lost money on the car, but eventually we were able to sell it. So the mashed-up Mustang lives on. A happy ending.

"Ferrari Fix"

The Ferrari F40, as it looked upon arrival. *COURTESY OF JOHN KRUSE.*

Like I've said before, "go big or go home" just ain't enough for me. Go biggest, go baddest, go raddest . . . *that's* when I make the big score. That's when I feel like I'm living. That's when I'm living the true Gas Monkey lifestyle that this brand represents.

When someone's faced with a completely wrecked Ferrari F40—a million-dollar supercar that was deemed "totaled" by the insurance company, as well as "totaled" and "unfixable" by Ferrari itself—any normal human being would run for the hills.

I'm not a normal human being.

First of all, Ferraris are built to incredibly precise specs. We'd need all kinds of expert help in bending and reshaping that frame to get it within three millimeters of where it was supposed to be or the car would be a complete failure. Second, there's pretty much only one place to buy Ferrari parts, and that's from Ferrari. Ferrari is an incredible company with decades and decades of reputation and history to uphold. They don't take too kindly to anyone messing with one of their vehicles—especially a bunch of cocky hot rodders. So we knew we'd have to start this rebuild on the down low. As soon as we got it in, we worked with some other Ferrari folks in the Dallas area and started ordering parts in a sort of quiet way that wouldn't draw attention to ourselves.

I'm glad we did, because as soon as Ferrari got word that we were attempting to rebuild an F40 that they'd deemed unbuildable, they shut down all related parts shipments to the U.S. for three weeks just to try to stop us in our tracks!

Aaron wasn't afraid to tackle this project at all. He loves a new challenge, and getting to learn all about a Ferrari F40 from the inside out was the chance of a lifetime. We'd be calling in some Ferrari experts as well, and he'd get to watch and learn from them. I think I said something about paying for Aaron's educa-

The F40 with a bit more work to go. *COURTESY OF DISCOVERY COMMUNICATIONS.*

tion for the last eleven years? That Aaron went to the school of Gas Monkey Garage? Yeah. This was a prime example.

The fact is, even when it comes to a Ferrari supercar, Aaron was cocky enough to think that he could build it better than it was originally. Me? I agreed with him!

I just never imagined quite how big the cost was going to be. I thought I might sink $100,000 into that car. By the time we were less than halfway finished, I realized I was off by more than fifty percent.

The Ferrari took us more than a month to build, and in that time I decided to flip cars that weren't just safe little $2,000- or $3,000-profit-type cars, but to take some big crazy gambles—just like I was doing on the Ferrari.

Guess what? Going badder and radder paid off for me, like it always does.

Who could forget that whacked-out Grateful Dead tour van I found, covered in stickers and filled with Grateful Dead memorabilia? I flipped that for a $40,500 profit. I flipped a '58 Corvette for a $10,998 profit. I picked up a '66 Porsche and made $12,000 on that. Then I made $12,485 on a 1917 REO. That's like seventy-five grand in less than a month!

With all of my big risks paying off, I decided that we should take another risk and paint the Ferrari black—just to fly in the face of everybody who said this build couldn't be done.

When all was said and done, we proved that Gas Monkey had the skills and wherewithal to tackle a seemingly impossible project—and we made ourselves a boatload of cash in the process. A hundred-grand profit on a single, forty-eight-day build.

Dennis Collins bought that Ferrari, and a little while later, just for fun, we got ahold of two other Ferrari F40s to do a little side-by-side comparison. These were pristine Ferraris with only a couple-hundred miles on them. We took them on head-to-head in terms of speed, acceleration, braking . . . and our rebuilt Ferrari blew those cars away. Aaron really had improved upon the original.

After the episodes aired in June of 2013, Dennis decided it was time to sell that car, and he took it to Barrett-Jackson, the biggest classic-car auctioneers in existence. Guess who bought it at

The finished product, out for a heart-racing test run. *COURTESY OF DISCOVERY COMMUNICATIONS.*

auction? Reggie Jackson. Mr. October himself, the baseball Hall of Famer, picked it up for $742,500. Sadly, that meant Dennis lost some money on the car. But Reggie Jackson got himself a bargain, and one heck of a rad ride. He drove it for a year or so, and then put it on the auction block again. Want to know why? Because the car was too fast for him, he said!

In the end, there were plenty of Ferrari purists who were pissed that we'd messed with one of their cars. But there was another faction of Ferrari enthusiasts who cheered us on. One of them even wrote a nice review in a magazine, saying (and I'm paraphrasing here): "Like it or not, Gas Monkey Garage rebuilt a better Ferrari. They took a totaled F40 that Ferrari said themselves could not be rebuilt and they made it great. So too bad."

"DALE JR.'S SICK NOMAD"

Aaron and me, with the finished Diet Mountain Dew–colored Nomad.
COURTESY OF DISCOVERY COMMUNICATIONS.

If I had any doubt that Gas Monkey Garage was turning into a huge success, it was pretty much put to rest with the very first episode of season four. It was awesome finding out that Dale Earnhardt Jr. was a fan of the show, and a real big boost to think we'd be doing a build for such a well-known celebrity client—with a corporate backer in Diet Mountain Dew.

It's funny to think that gigantic opportunity was almost lost because Christie didn't believe that Dale Earnhardt Jr. would actually be calling our little garage down in Dallas. She kept hanging up on him! The fact is, Gas Monkey was stepping up in the world—and was catching the attention of some very big fish. Before long we'd wind up building a Camaro for Sonic, and doing a pink Cadillac for Hard Rock Hotel & Casino. Those builds would bring Gas Monkey even more notoriety, plus, you know what's really fun about those big-client corporate builds?

Delivering to the happy customer himself, Dale Jr. *COURTESY OF DISCOVERY COMMUNICATIONS.*

They make a lot of money. Just check out the numbers on Dale's Nomad alone. I paid $26,000 for the car. I invested $22,000 in parts and labor. Then Aaron and I spent $350 for gas, food, and lodging on our road trip to North Carolina to deliver that car to Dale, which means all in all, I walked away with a cool $26,650 profit.

We did it in ten days. That's more than $2,600 in profit per day!

Find me another business that pays that well, and . . . hell, I just might buy it.

"BANDIT CAR"

That's me playing the part of the Bandit, mustache and all, in the finished Trans Am. *COURTESY OF RICHARD RAWLINGS.*

Who in the world would be crazy enough to pay me $70,000 to build them a rally car, throw in a $40,000 bonus if I could get it done in six days, and then bet me another $25,000 that I couldn't get a certain movie star's signature on that car in that six-day time frame as well? None other than my buddy Jay Riecke—the very same buddy who bet Dennis Collins and me that we couldn't beat the world record in the Cannonball Run.

As you know, I earned a payday on that Cannonball Run bet back before *Fast N' Loud* was on the air. There was no way I was going to lose this one!

The real bonus for me, though, was getting a chance to meet Burt Reynolds, the man himself, the legend, the star of so many favorite films from my younger days including *Smokey and the Bandit*, *The Cannonball Run*, and *Hooper*. Did you know that Burt did his own stunts on all of his movies? He's one badass individual.

The fact he's still walking around at all after all the stress he put on his body and all the running around he did at the peak of his stardom is a freakin' miracle. He's a legend, and that's all there is to it.

Something cool happened off camera at Burt's house, too. He waited until filming was done to do it, but it's a moment that I'll never, ever forget. We spent some time talking to Mr. Reynolds about Gas Monkey, and how much time Aaron and I had put into building the brand. I also told him about beating the world record in the Cannonball Run, and I showed him my tattoo commemorating that feat.

Just before we left, Burt disappeared into another room and came back with something in his hands.

"Do you know what this is?" he asked me.

"Yes, sir, I think just about every guy on the planet knows what that is," I said.

The *Bandit* hat Burt Reynolds gave to me, proudly displayed on a shelf in my office. PHOTO BY MARK DAGOSTINO.

He was carrying his black cowboy hat from *Smokey and the Bandit II.*

"Well, I want you to have it," he said.

"Sir, I couldn't possibly take that," I told him.

"Well, you are gonna take it. So here," he said as he handed it to me. "You're carrying the torch."

I don't like to admit this sort of thing, but my eyes actually welled up with tears. I couldn't believe it. I thanked him over and over for the gift. It means the world to me. I keep it up on a shelf behind my desk in my office now, so I see it every time I walk into that room. It's right there over my shoulder, reminding me how far back my love of cars goes, and reminding me just how far this journey has brought me.

"Aaron's Falcon Race Car"

Something we've wanted to do here at Gas Monkey from the very beginning was to build purpose-driven cars—race cars and rally cars for specific races and rallies, whether it's the Texas Mile or a hill climb or another Gumball 3000. Apparently our TV fans like that idea, because one of the most talked-about cars we've ever done is Aaron's Falcon.

Racing that car to the top of Pike's Peak in the annual Pike's Peak International Hill Climb was a big moment for Aaron. I asked him about it recently, and he said, "That's the coolest thing I've ever done. It's something I've wanted to do since I was a little kid. I never thought I'd have the opportunity. I'll be going back to race it again for the rest of my life, as long as I can afford it."

He'll keep taking that car back, too. He's done a bunch of modifications to it since he first got his hands on it.

"My primary goal was simple," he says. "There's a lot of

Aaron's Falcon. *COURTESY OF DISCOVERY COMMUNICATIONS.*

people who race for years or decades and don't summit. There's way better drivers, more experienced drivers than me, that don't summit. So all I wanted to do in my first race was I wanted to finish. I wanted to have a time and not have a 'DNF.' I just wanted to see the top."

He did. In fact, he did it in under thirteen minutes. The fans loved it. Everyone in the shop loved it, too. Most of the guys weren't able to travel to Colorado to see the race live. We had too much work that needed to get done. So when that episode aired, a bunch of them got together and watched it on TV. It was a real thrill for every one of us at Gas Monkey to watch Aaron summit in that car. It felt like a real achievement. For everybody.

Aaron cried when he reached the top, only none of us knew that until the episode aired. We saw it on TV, just like all of our fans did. We got all kinds of Facebook comments and tweets

Aaron's Falcon, all ready to race. *COURTESY OF DISCOVERY COMMUNICATIONS.*

The pace truck in all its glory. *COURTESY OF DISCOVERY COMMUNICATIONS.*

from fans saying they cried right along with him. It's an emo-
tional thing to watch somebody achieve a lifelong dream.

"It's only going to get better," Aaron insists. "There are plenty
of things coming for that car. I'm hoping they'll be all done for
next year's race."

Besides being a big moment for Aaron, that race provided a big
moment for Gas Monkey: Building the Pike's Peak Pace Truck for
that event was something everyone at the shop was proud of, and
one more chance to show the world what Gas Monkey is all about.

"Holy Grail Firebirds"

Nearly two years after we'd gotten this crazy TV show off the
ground, we found not one but two cars that knocked us for a
loop: the first and second 1967 Pontiac Firebirds ever made.

Chuck Aleksinas, a former University of Kentucky and NBA
basketball player, had them sitting in his garage. He'd purchased
them online a while back and then never got around to restoring 'em. He wasn't really planning on selling them at that point.
Wasn't planning on doing much with them at all—until we
found out about those unbelievably rare automobiles and came
knocking on his door.

Once we got a firsthand look at them, we jumped on the
Interwebs and did some serious research into those Firebirds and
their history. Here's what we found out. When Pontiac intro-
duced the Firebird in 1967, it promoted five distinct models and
called them the "Firebird Magnificent Five." The engines and
body styles were the main differences between them. The serial
numbers on the cars we found were 1000001 and 1000002.
These cars also had trim tags clearly marking car number one

with SHOW1 and car number two with SHOW4. So the two Firebirds we'd located were two of the original cars that made up the
Magnificent Five. Show cars don't often survive. They're taken
around the country and shown off to get dealers and car enthusiasts excited about the new line, and then they disappear. Just
like in the art world or antique world or anywhere else, having
the "first" of anything is a big, big deal to collectors. We knew we
really had something.

When I was able to buy these cars for a grand total of $70,000

The Firebirds as we
found them—number
one (above) . . .
*COURTESY OF DISCOVERY
COMMUNICATIONS.*

. . . and number two. *COURTESY OF
DISCOVERY COMMUNICATIONS.*

combined, I just about soiled my pants. Our research indicated that these cars—fully restored—could be worth in the ballpark of a million dollars. A *million dollars!*

Then I found a potential buyer who wouldn't balk at that potential price tag if we could get both of those cars restored to Concours standards in sixty days. (Concours standards are those that reflect "best original condition," and are generally regarded as the highest standards in the business when it comes to classic-car restoration.)

This was the ultimate chance to prove ourselves in the restoration world, and the ultimate chance to make a gigantic amount of money in the process. I had some cash in the bank. I felt like Gas Monkey was on sure ground. Now, once again, I knew I'd be risking almost everything to try to get these cars built to exacting standards in just two months—the sort of build that people might otherwise expect to take years.

A lot of change happened at Gas Monkey during those two months, too. First of all, Aaron is not a Concours restoration expert, which he readily admits, so I hired Jason Aker and threw him right into the mix on one of the toughest builds we'd ever had. I knew we needed his level of professionalism to get this job to the finish line.

I also knew that two of the guys on my team had been proving themselves somewhat less than professional with each passing build: Tom and Jordan. I haven't spoken of them in this book because, well, why would I? They're not here anymore. I canned 'em right in the middle of this build and I haven't looked back. No one should feel sorry for them, either. Those clowns managed to land their own Discovery spin-off show called *Misfit Garage*.

In the end, the newly slimmed-down Gas Monkey team created what I felt were two flawless examples of Firebirds: car number one, the convertible in red, and car number two, the HO (so named for its High Output engine) in silver.

I sold those cars for $650,000 for the pair. It wasn't the million bucks I'd hoped for, but it was still the biggest single payday in Gas Monkey history. After investing $340,000 and two months of our time in the restoration, I walked away with more than $200,000 in profit—and a whole new set of bragging rights for Gas Monkey Garage.

Through the years, I've actually owned quite a few number-one cars. I've had '69 Firebird number one. I've had '68 Camaro number one. I've had '71 Barracuda number one. There are other

The completed Holy Grail of Firebirds, revealed in front of a live audience. *COURTESY OF DISCOVERY COMMUNICATIONS.*

number ones out there. If you search hard enough, you can find them. And if you do, I'd encourage you to bring 'em on down to Dallas. Because the fact of the matter—and the truth of the matter—is that Gas Monkey Garage is capable of restoring any car to its Concours-standard glory.

For the right price, of course.

There are so many other episodes that could qualify as favorites. I mean, a lot of fans went crazy for the "Frankensteined Ford," where we pieced together a '68 Ford F-100. What some viewers don't realize is that after the show aired, K.C. went out and bought that truck back for himself. He's always in here making

Aaron and me with the completed "Bad-Ass Bronco." *COURTESY OF RICHARD RAWLINGS.*

modifications to that truck on his own time, and it's even cooler now than when we got it finished on TV.

There are those of you who love the "Bad-Ass Bronco," or the "Chopped and Dropped Model A," or that crazy VW Shorty we did, where we fixed up a VW bus that'd had its whole middle chopped right out. I can't say any of you are right or wrong. I hope every episode we do makes for great viewing.

I will tell you this, though: one of the builds that pretty much every guy in the shop wishes could be theirs came right as I was writing this book. It's the truck I'd like to keep for myself. It's a truck that Aaron wishes was his own. It's a build that everybody poured their heart and soul into because we had total freedom and creativity to make it the best truck we possibly could, and to build it on a really fat budget, too.

I'm talking about the 1976 Chevy Silverado C10 pickup that we dropped for the big Cattle Baron's Ball in season five. It's rare that any one build we do would line up with everybody's tastes. We all like what we like, you know? But I truly think that everybody is cool with that green-and-white color scheme. I think everybody is cool with the trim. I think everybody likes the rich, totally custom-stitched leather interior. They like the stance. They like the wheels. If they had the time and opportunity and money, I think everybody at Gas Monkey would have built it for themselves.

The funny thing is, it didn't go for anywhere near what it should have at that charity auction. I guess if we'd wanted to appeal to the good ol' boys and Texas oil millionaires, we probably should have lifted the truck instead of dropping it. Maybe put some horns on the front. Some lights on the top? But the C10 represents everything that's great about Gas Monkey. We're not your standard builders. We're not out to do something that everybody

The finished Silverado C10, just before we took it to auction. *COURTESY OF DISCOVERY COMMUNICATIONS.*

else can do. We're about creating a unique style, and that truck has it. We're about performance, and that truck *performs*. We're about making a statement, and that truck makes a statement anywhere it rolls. I know, because I bought it back! Dennis did me a favor and bought it at the auction, then I bought it right back from him. I wasn't gonna let someone get a big bargain on that truck. I figured I'd rather drive it myself, and maybe eventually sell it to someone for what it's truly worth—and that truck is easily worth upwards of $150,000. My estimate for auction purposes was $250,000, and that was no exaggeration.

When we finally took it to Barrett-Jackson to get a fair price for it in early 2015, it sold for an extremely disappointing forty grand. Forty grand! I hope that buyer realizes what he's got, and appreciates it for the craftsmanship and care that went into every single detail.

When it comes to favorite shows and favorite builds, Aaron's take is that the shows don't really matter. "We make tangible goods," Aaron says. "You can touch them. We're responsible for them. We're responsible for their quality. Twenty years from now, if the show is done and gone and no one remembers it, that car we made will still have everyone's name on it."

I disagree. I think the television we're making is making a big difference, too. I think we're touching people's lives, and in-spiring them, and giving them some great entertainment. That has as much or more value than the individual vehicles to me. I think we're showing off just how much fun it can be to live a hot-rodding life.

Some of my favorite times at Gas Monkey Garage are after the sun's down, when the cameras are gone, and some of the guys in the shop are adding unique touches to their own personal vehi-cles after hours. Standing there drinking Miller Lite, shooting the s—t, talking about tires and wheels and brakes and pistons and viscosity—those are good times.

The hot-rod business can be anybody's business. Rich guys, poor guys, they can all play on the same field. They really can. At the end of the day, everybody's hot rod is different. It's a different idea. Surely you get into cliques. You might get a '69 Chevelle fan and they've got a couple of buddies who are into those same cars, too. But you can literally be a hot-rod guy and have anything, from a Metropolitan convertible from the fifties to a thirties car to a muscle car. You name it. Hell, I've seen box trucks get turned into hot rods. It doesn't matter. I've seen an actual trash truck made into a limo. People were sitting in the back! It really has no boundaries.

I think all of that stuff's fun. I mean, if the average person out there watches the show and gets excited, I think it opens up possibilities in people's minds. People might think, "You know what? I could go out and find an old car to have, and it doesn't have to cost me $50,000." That's really the part I enjoy the most. I want people to watch my show and not be deflated. I want them to be energized and think, "Wow! I can do that, too!" They can have something like that, or flip cars like that, or maybe they can build something. Whatever. I mean, hell, I've got a $200 Pinto that I keep in storage. I think it's freakin' cool. Or at least it *could* be cool. You could do a bazillion different things with a $200 Pinto.

I keep that car as testament to a really important principle in the life of a Gas Monkey. I keep it to remind myself, "You know what? One day things might not be that awesome or they might not work the way we want them to—but I can still have a hot rod."

It may be a Pinto, but how many Pintos do you see on the road anymore? If I brought that thing out of storage right now, people would see it driving down the street and go, "Hey! Look at that! It's a Pinto!"

And you know what? That's pretty cool.

LET'S HEAR IT FOR OUR FANS!

I want to stop right here to say a little something about our Gas Monkey and *Fast N' Loud* fans.

Part two of the Bandit Car build grabbed our highest ratings ever. More than 3.1 million people tuned in, and that's not counting all the people who've watched it on repeats.

That episode garnered so much attention, it really put a spotlight on just how much our fan base had grown since the show first started back in June of 2012. Eighteen months after we'd first begun, we had a certified hit on our hands. *Fast N' Loud* became the number-one car-related show in America, and we've been rockin' and rollin' ever since.

Whooo!

And it's all thanks to you! The fans!

You guys blow up our social-media feeds every single time an episode airs, and you send us all kinds of great feedback and questions, too. We love it. Every Tuesday morning after a new episode airs, I come into the office to find five or six hundred messages waiting for me. It floors me every time that y'all keep watching and wanting to know more.

I wish I could answer every one of those messages personally, I really do. And I answer a lot of them myself. But there isn't enough time to flip cars, build cars, keep building the Gas Monkey empire, and respond to hundreds of e-mails in a day, too. So thankfully I have some great people around here who help me out and who take care of the fans as best we can. I love you guys, and I want to make sure every single one of your messages gets a response. It's not easy to keep up with y'all!

Just saying hello to fans who recognize me when I'm out and about in my regular life takes more time than I ever could have imagined. I love meeting fans and saying hi and taking selfies with all you Gas Monkey maniacs. But again, it's hard to keep up! I mean, I go to the grocery store and people are like, "Oh, can I take my picture with you?" I'm like, "I'm buying toilet paper, man." It's cool, though. Without the fans, I don't know where we'd be.

My fans aren't just local, either. *Fast N' Loud* now airs in over 150 countries worldwide. We're a huge hit around the globe! The magic is that we're a show all about cars and motorcycles. That s—t translates. Cars and motorcycles and beer and women translate in any country. Every day we have people from out of the country who stop by our shop here in Dallas to buy T-shirts and get a glimpse of the garage they've seen on TV. It's awesome. It really is. People are building their whole vacations around coming here, and I couldn't be more grateful.

From the very beginning, I wanted Gas Monkey to be a worldwide lifestyle brand. I specifically wanted the show to be fun and accessible in a way that even people who can't afford to have a hot rod would be able to watch the show and have a good time and enjoy it. So far, that's exactly what seems to be happening.

Gas Monkey is not a garage so much as it is an attitude. It's what I want and how I want it. Whether it's cool or not, it's cool to me. It's not about being the baddest guy out there and, "Oh, look at me, I'm a tough guy," or whatever.

I mean, the average Joe can be a Gas Monkey. If he's a guy wrenching on his car in his garage and he takes it out and people laugh at him because it's a green '70 Chevy Nova four-door. . . . Well, you know what? He f—king worked on that. He's a Gas Monkey!

Being a Gas Monkey is about getting out there and getting dirty with your hands, having a good time, and enjoying the fruits of that labor. At any car show, I'm more impressed with the guy who shows up hand-polishing his fenders on the beast he spent

Me, getting my own hands dirty in the shop. *COURTESY OF DISCOVERY COMMUNICATIONS.*

years building in his backyard than I am the rich dude who sinks $400,000 into a car just because he can. I hold that little guy in higher regard because he put his own blood, sweat, and beers into that build. I'll never lose sight of that, even as Gas Monkey grows and tackles bigger, more expensive vehicles for all sorts of rich clients and corporate sponsors. Staying connected to the real fans, the gearheads, the folks who might not even own a hot rod of any kind but who feel that they're Gas Monkeys at heart—that's gratifying, and honestly it's one of the best parts about being involved in this TV show.

Connecting with all of the fans on social media has been a trip, too. We hit three hundred thousand Facebook fans by December of our very first year on the air—and the show had only started that June! We've added millions of fans on Facebook since then. When I finally got around to putting up a personal Facebook page, I went from zero to seventy thousand fans in the first week. Christie has hundreds of thousands of Facebook fans herself. (And she's done it without posting many naked foot photos!)

We've been putting a big push into Twitter, too. I live-tweet during every new episode. Getting everyone's feedback in real time on Twitter is awesome. I almost feel like we're at one big watch party where I'm sitting in everyone's living room with them watching *Fast N' Loud* together. I love all the pictures people send, too, like, "Hey, here's me and my dog watching the show!" or "We all have our shirts on and our Miller Lites!" How cool is it that people are sitting around their living rooms with buckets of Miller Lite, and the guys out there are enlisting their wives and children to be their beer assistants while they watch our show? (I'm sure some of you ladies have your guys be your beer assistants, too. Don't go writing me nasty letters, now, y'hear?)

We've also got a lot of what we like to call "superfans" of the show who just love to take to Twitter and get into fights about stuff. It's pretty funny. I especially like the super girl fans. Some of y'all are hardcore. There have been groups of girls who all got together from different parts of the country and came here on vacation, hell-bent on meeting me. I've even gone out for drinks with some of 'em! Why the hell not, right?

I remember there was this one lady coming in from Scotland and she was about to cancel her trip because her friend bailed on her or something, but then she tweeted about it and one of the other Gas Monkey Girls (which is what we call our female super-fans) tweeted her and insisted, "No, come on down! I'll meet you down there!"

That girl wasn't even in Texas. She came all the way over from Louisiana or something. I can't remember. They met here, hung out all weekend, and took all sorts of pictures.

We've even had a Gas Monkey fan marriage. No joke!

We did an episode that included a Pantera that we just couldn't sell, and that episode drew all kinds of hatred of Panteras in our Facebook comments. People were just bashing Panteras for being terrible cars, and this one girl made a comment that came to the defense of the car. Two comments down, some guy responded, "She knows what she's talking about," and right then and there, love was born.

Those two with their mutual affection for Panteras wound up commenting back and forth, and then took their talk offline. Then they met in person. They lived on opposite sides of the country, but they sought each other out, and next thing we heard they were thanking us for allowing them to connect, and they were getting married!

Our show touches people in all sorts of ways. We used to get a whole bunch of messages from this one guy, and we would reply to him, of course, like we do to everybody. But then a couple of months went by without him writing. It turned out he had passed away, and the next thing we knew, his wife sent us a note, saying, "I just want to thank you guys so much. You made my husband's week every time you guys replied to him. He was just so excited and it kept him in good spirits while he was sick."

To me, that's what this is all about. If we're connecting with our fans, then our show and what we're doing here matters. It's making people happy. There ain't nothing better than that.

Aaron has his own thoughts on why the show is popular with men and women, young and old: to him, it all comes down to the cars. "How many people do you know that don't own cars? Probably next to none. And of those people, how many people's parents own cars? How many people remember going on family vacations in cars? How many people have a percentage of their childhood memories based around a car, in a car, traveling and what have you? What's been the biggest symbol of freedom in your entire life? I would venture to guess it was when you turned sixteen and got your driver's license. Who doesn't remember the feeling that they had the first time they brought their first car home? It's almost better than the first kiss you ever had."

Aaron's right, in that the love of cars is a deep-rooted part of the American experience. There hadn't really been any shows that captured that on the television landscape until *Fast N' Loud* came along.

I think there's another key to why we have so many loyal fans, though: I think people really like to watch people make a living doing something they love. Whatever that thing is, if you're passionate about it, and you're into it, and you're giving it your all, it just fires people up to see it unfold.

And you know what? I want nothing more than for every

Gas Monkey fan there is to go out there and make a living doing something they love, too. It's one of the greatest feelings on earth to wake up every morning knowing that you get to do what you love. At the very least, you can turn a passion into some extra income on the side and have a whole lot of fun while you're at it. And if your passion is cars, then maybe you ought to think about getting into the car-flipping business yourself. I'll spend the last part of this book sharing some tricks of the trade about flipping cars, all in the hopes that some of you fans will go out there and tackle it and make a killing like I have.

I know you fans are crazy about this show, and I also know that some of you people are just plain crazy! For example, I dig all the cards and letters and random gifts that you send (like hand-painted portraits of my face). And if you're pretty, I appreciate the nude photos and all the sexual wish-list notes, too. Just don't go all stalker on me, okay? That stuff gets weird. It's not exactly comfortable having to call the cops on someone who won't leave the premises—or to sic Christie on 'em, either. (Like I said, Christie's tough. She'll cut a bitch!)

Those of you who want to go out and get Gas Monkey tattoos, though, I'm all for it. Why not? It's free advertising for me! We've seen pictures sent in from one guy who got a huge Gas Monkey tattoo on the back of his bald head. Another had his complete forearm done in nothing but our logo. There was a guy at the airport who came running up to me and pulled half of the backside of his pants down to show me the Gas Monkey he'd tattooed on his butt cheek. Then there was a guy who got Aaron's face tattooed on his ass. Ha!

Okay, that one's kind of my fault. We threw a big viewing party one night, and a fan drew a picture of Aaron's face and put it on Facebook. I took one look at it and issued a challenge to everyone at the party: "I'll pay someone $1,000 if they get this tattooed on their butt."

This one guy was like, "I'll do it."

I was shocked. I was like, "Really?"

We e-mailed the image to the tattoo place next door, they printed it out, the guy went over there, and sure enough now he's got a big old picture of Aaron's face and beard on his butt cheek.

Another woman who contacted us through Twitter has a tattoo of a monkey with a goatee on it (which is supposed to look like me), tattooed along with the phrase GET YOU SOME OF THAT! She was actually scheduled to visit the garage before this book came out, and she said she wanted me to sign my autograph near her tattoo so she could get my signature tattooed on herself, too. I was like, "Fine. Whatever you're into!"

The fact is, we get lots of messages now from people asking if it's okay to use our logo for a tattoo. My response is, "Have fun, but please don't butcher it."

There was one guy in an airport who came running after me: "Roger! Roger! Roger Rawlings!" he yelled. He got my first name wrong, and yet he had this big Gas Monkey tattoo on his arm. I posed for a picture with the guy anyway, because you know what? It doesn't matter if everyone knows my name. What matters is that they know Gas Monkey Garage, that they know our reputation, and that they get excited about the show.

The Gas Monkey fandom has grown so big, I've been able to land spokesman deals with two of my favorite companies on the planet: Miller Lite and Dodge. My face is on one of the biggest Miller Lite billboards in the whole Dallas area now. I can see it from where I live, which means there's a giant me staring back at me through my window. Let that s—t sink in and mess with your head for a minute. Some of my employees complain that they have to see my face ten times on their morning commute before they've even gotten in to work. I tell 'em everyone should be so lucky!

In addition to providing me a 392-Hemi Challenger to drive around in, Dodge jumped in and helped us throw a massive event here in Dallas, complete with a concert by Mötley Crüe. Just about the only music I listen to is hop-hop and hair bands, so to be involved in a Mötley Crüe show was a dream come true.

The list of cool stuff goes on and on. I mean, even Havoline's a sponsor for the show, and they've jumped in on some really cool giveaways for the fans, including an amazing green Camaro I picked up at auction that drew more than 150,000 entries on Facebook.

The fact that our fandom keeps growing means we gain new sponsors all the time, which allows us to do more cool contests and events for the fans, which earns us more fans as a result. It's an unbelievably cool cycle to get caught up in, and all I can say is, "Thank you!"

THE GROWING GAS MONKEY EMPIRE

Fast N' Loud has taken off so much that Gas Monkey Garage is in the process of expanding—big-time. I just bought the property on each side of us here in Dallas, and we'll be blasting through the walls next door by the time this book hits the stands. One of the things I'm thinking about adding to the customer experience here is a sort of car museum—a place to show off my personal car collection and some of Gas Monkey's favorite builds from the past. I would love to let our fans see some of these cars up close. I know they would love it.

I keep a stash of cars and bikes for myself over in one corner of the shop as it is. Among my favorites, I've got a kick-ass replica of the Porsche James Dean was driving when he died. It's been listed among the finest replica cars ever built, and you'd know why if you got behind the wheel.

My very recognizable '68 Shelby with the roll bar, which we

fashioned to look just like a car that briefly appeared in the movie *The Thomas Crown Affair*, is there, too. My green Harley is there, the one I said I would never part with, right next to a crazy-looking chopper I bought online one day.

My James Dean–style Porsche replica. *PHOTO BY MARK DAGOSTINO.*

There's an old-fashioned 1929 Packard, with a big long front and ornate details—a fantasy car from another era that I always dreamed of owning. I also recently picked up a gold Ferrari 308—that's the Magnum P.I.–style Ferrari. Those cars have come way down in price in recent years, and I managed to find a gold one—one of only six gold ones ever built.

My 1929 Packard. An old-school dream car of mine . . . that's now mine! *PHOTO BY MARK DAGOSTINO.*

On the nostalgia front, I repurchased the first car that Gas Monkey ever built: a dropped '52 Fleetline with an old patina finish. I also bought back an incredible motorcycle that Aaron built entirely by hand, piece by piece, three or four years before we landed the TV show. I like having those old builds around to remind me of what we're capable of, and just how far we've come.

I take most of those cars and bikes out and drive 'em, too. I don't like treating cars strictly as something to look at. Cars are meant to be driven. A few chips in the paint, some road wear, and the smell

The first car that Gas Monkey Garage ever built. *COURTESY OF RICHARD RAWLINGS.*

of burned-out tires make any car—even a really expensive car, like a Ferrari—that much cooler.

The way I see it, I've got my retirement in the corner of this garage. It's my retirement, but I get to have fun with it *now*. I get to look at it every day, and I get to drive it if I want to. That's a whole lot more fun than leaving my money in some bank.

The growth of *Fast N' Loud* has definitely allowed me a few perks in life. It's also allowed me to finally see through my early vision of turning Gas Monkey Garage into a brand that's recognized all over the country and all over the world. In addition to our little shop at the front of the garage, we've got Gas Monkey apparel and merchandise showing up in major clothing and department stores all over the place these days. And we're just getting started! We've seen the first Gas Monkey spin-off show in *Misfit Garage*, and a few episodes of yet another Gas Monkey spin-off called *Demolition Theater*—where me and some of the other Gas Monkeys sit around watching crazy videos of people crashing cars and doing crazy stunts and just laugh our asses off. It's crazy, and I love it. But it's *still* just the beginning.

I created a concept for a scripted TV show, based on a garage that's a little something like GMG, full of all sorts of characters that are reminiscent of Daphne and Christie and everybody, and we sold it to actor Vince Vaughn's production company. Heck, I could be the first so-called reality star to turn myself into a Hollywood mogul if I keep this up. And you know what? I plan to keep it up! It's fun, and it's making me money. So why not?

We also broke into the NHRA drag-racing world during our fifth season on the air, buying our dragster and jumping in the game without corporate sponsors, which is just unheard of. We didn't come out on top in our first race, but Gas Monkey Racing is going to take that sport by storm in the year ahead. Just you watch!

Just a small sampling of our Gas Monkey apparel and merchandise. *COURTESY OF RICHARD RAWLINGS.*

Right here in Dallas we've already opened the first-ever Gas Monkey Bar N' Grill, a restaurant concept that's far exceeded even my wildest expectations of what a successful restaurant can be. I saw the potential of that property sitting right off a major freeway, so I bought it and I assembled a team to draw up designs and set up a kitchen, and off we went. The spot we're in used to be a restaurant called Firewater Bar & Grill, and what we did was the same sort of thing we do with cars: we gave it the Gas Monkey once-over. We took that old place and turned it into something entirely new and spectacular. It's rustic-looking now, with reclaimed wood and all sorts of old rusty metal signs and things on the wall. We built a stage, and a VIP lounge upstairs, all overlooking that man-made pond with fire elements that shoot up from the water. It's rad! Our kitchen cranks out great food, our bartenders are the best in the business, and we treat our customers like customers should be treated. After all, our customers are Gas Monkey fans, and those fans deserve all the thanks and praise in the world.

You find a lot of so-called "celebrity" restaurants in the world, and yet the "celebrity" is never in the place. They just lend their name to it and sit back to collect a paycheck. It's not like that with GMBG. Me and the other Gas Monkeys are in there whenever we get a chance. I'll admit, sometimes it's not easy. If I show up when the place is hopping, it might take me an hour just to make my way to a table 'cause everybody wants to shake my hand and take a selfie with their cell phone, you know? But I take every one of those pictures, and I thank every person I come in contact with, because if it wasn't for them, I wouldn't be living any of this dream right now.

What amazes me is how many of our fans fly all the way to Dallas from Germany or Sweden and plan their entire vacations

Gas Monkey
Bar N' Grill
getting ready
to open (above) . . .
*COURTESY OF
RICHARD RAWLINGS.*

. . . and open! A view from
just offstage, and inside the
main dining room. *COURTESY
OF RICHARD RAWLINGS.*

around a visit to Gas Monkey Garage and Gas Monkey Bar N' Grill. I love it! I've been to London a couple of times since the show started airing in Europe and I can barely make it through the airport. People go nuts.

My plan is to take the Gas Monkey Bar N' Grill phenomenon and spread it all over the country, and all over the world. We've already got a mini location at Dallas–Fort Worth International Airport, and we're opening a spot in Southern California. Heck, we could open this fun, laid-back atmosphere of a restaurant in all sorts of island and vacation destinations, too, and then I'd have to go inspect each one. (Tough job, but somebody's gotta do it!) I'm fixing to open a bunch of Gas Monkey Roadhouses, too, like old-fashioned drive-ins, but ones that invite you to come in and have a beer.

Speaking of beer, that makes me think of bars—and as I'm writing this book, Gas Monkey is getting ready to invade bars all over the planet with Gas Monkey–branded tequila. We're talking thousands and thousands of cases of the highest-quality tequila imaginable, kicked up a notch in true Gas Monkey style with a spice that'll knock your socks off. We've already got a distribution deal in place with Bacardi, and I expect that all of you spring breakers out there have already tried it by now.

We've got a Gas Monkey concert venue now, too. Just up the road from the Bar N' Grill I bought a 2500-seat live-music mecca and turned it into Gas Monkey Live! We've welcomed all kinds of bands there already, from Reel Big Fish and Los Lonely Boys, to Social Distortion and one of my personal favorites, Mötley Crüe.

The Gas Monkey empire keeps growing, and the reality of seeing that dream of mine come true is more rad than I ever imagined it could be. Considering the fact that *Fast N' Loud*'s only been on the air since 2012, just imagine what you'll see from Gas Monkey Garage in another year, or two years, or three years?

All I can say is hold on to your socks, pal.

And to think it all goes back to my love of cars, and my fast-thinking, cash-driven desire to flip cars that goes all the way back to when I was sixteen years old. It turns out that flipping cars was a pretty good little business to get into. And that's what the next section of this book is all about.

PART THREE

FLIPPING OUT

Cash money! That's my smile after making a quick sale. *COURTESY OF DISCOVERY COMMUNICATIONS.*

THE SECRETS OF MY SUCCESS

A lot of people ask me, "Do you think it's possible for the average Joe to get out there and flip some cars?"

My answer: absolutely! That's what I built my business on! I think it's possible to make a very good living doing what you love, whatever that love may be. It's like anything else in life, though: when it comes to buying and selling, you have to know when to hold 'em and know when to fold 'em.

On *Fast N' Loud* we focus pretty heavily on the builds, the big-money, super-challenging projects that come our way. But a lot of times it's those B-stories, the cars I'm flipping, that actually provide the money that keeps Gas Monkey Garage solvent. When I lose money on a build, I make up for it by flipping cars. Sometimes it's with a crazy-lucky flip, like the Mustang I paid $65,000 for and sold for $125,000. But more often than not it's the multiple cars we're flipping for a profit of $2,000, $4,000, $6,000, and so on.

I mean, I've got a Model A in the parking lot right now that we threw up on eBay. Last time I checked, it's at $9,500. I only paid $4,000 for it, and I didn't do anything to it! We bought it, had it delivered, took some pictures, and threw it on eBay. That's it.

I know what you're gonna say: "Well, people are just paying more for that car because they're buying it from Gas Monkey Garage, and they want the bragging rights to say it's a Gas Monkey car." Correct. From what I've figured, the Gas Monkey notoriety has increased our profits on a lot of cars by a good 20, maybe 25 percent. Even so, let's take that out of the equation. A $5,500 profit minus 25 percent is still $4,125 in profit. That's more than doubling my money. Minus the freight cost, it's still nearly double what I paid for it. You could be a bazillionaire with a Swiss bank account and not make that kind of a return on investment anywhere else.

When it comes to flipping cars, ever since day one my goal has been to beat the banks. So my basic rule of thumb is this: if I can make 15 percent on the investments I'm making, I'm doing pretty great.

If you're looking to double or triple your money with every car you buy and sell, you're going to be disappointed. Sometimes you get a bad one and you have to just get rid of it as fast as you can for whatever you can get. Sometimes you get a good one and you hold on to it and get as much money as you can. As long as you can make a profit on average, you're good. If not, you're doing something wrong. But it is definitely not a game in which you're going to double your money every time.

I sure as hell like the 300 and 400 percenters, don't get me wrong. I look for those wherever I can find 'em. But even with those, when you average them out against the break-evens and losses, that 15 percent number is the one I'm usually hitting.

Plus, I look at it like this: do you want your money sitting in the bank making 3 or 4 percent in your savings account, or even 1 percent in some checking account? Or would you like to at least take your old lady for a ride in a Bertone X1/9 that you paid $800 for and can sell for $1,200? *Get you some of that!* I mean, come on! It's your money. If you can make a little profit at the same time you're having fun, that's a whole lot more valuable to me than letting it sit in the bank.

By the way, those X1/9s are horrible cars, and they're selling for a lot more than $1,200 right now, so they're probably a lousy investment. Don't go buying a Bertone just because I mentioned it here, okay? But you get my point.

I think a lot of people struggle. They're in a job, they have a certain amount of income coming in, and yet they're struggling every month. They don't think outside the box. If cars are something you love, you could invest some time and effort and money into flipping and maybe make a little something here and there on the side. There's a learning curve to it, and it can definitely be risky, but for those who want to dedicate the time and intelligence it takes, I really do believe it's a great way to make some money.

If everything else went away, I know that I could make a living flipping cars for the rest of my life. Back when I was going through my divorce, right at the same time that Gas Monkey was going through a rough period just before the show started, I took off and rented a penthouse apartment at a high-rise in Dallas. The only income I had at that time was from flipping cars. Period. People thought I was some kind of a drug dealer or something because I lived in a great apartment and lounged around the pool most days.

The reality was I kept four or five cars at a time, and I flipped them. I had zero debt in my life, and zero bills except for rent and phone, so my overhead was low. That came from years of being disciplined with my money. Not only could I make ends meet, I could live like a king. All I did for a few months was sit in a swimming pool, drink margaritas, and flip cars.

My overhead was so low that if I flipped one car and made ten Gs, I didn't have to do any more work that month. And I didn't even have to work all that hard at the selling because I flipped the cars online.

I've pretty much been online since it started, and that is definitely one of the keys to my success. I was a very early adopter of eBay and then moved on to Craigslist and other listing sites. I use the newspaper and AutoTrader and *Hemmings Motor News* and stuff like that to place ads that say, "I buy cars." I personally don't ever advertise my cars for sale in those publications at all. I advertise to *buy* cars. That way people come to me when they've got something they need to get rid of, and chances are, if they're answering an ad that says, "I buy cars," they're doing it because they need the cash—and that probably means I'm going to get myself a bargain.

When it comes to selling, all of my cars are on eBay or Craigslist, and that's it.

Now, keeping your overhead low to zero while you do this is key, and not getting buried by the cost of some lousy car that you're never going to make money on is the other key. That's where a lot of people get caught.

One of the biggest pieces of advice I have to give in terms of making this business work is this: Don't get yourself trapped.

Don't decide that you've got $15,000 in it already and you've got to do all this fixing it up to get to a certain point before you can let it go, and blah, blah, blah.

Do your math!

If it's a loser when you bought it for fifteen grand, well, it's not going to be a winner when you've sunk a bunch of time and money and effort into it. Get rid of it then and there so your margin of loss is a lot less. Sell it to somebody else who wants a project car. There are plenty of those folks out there.

Cutting your losses before they get worse is hard for some people to do, and you need to assess whether or not you're one of those people. If you are one of those people who can't let go, learn that about yourself and find a way to change your behavior. If not, this business might not be for you.

Don't be that guy!

Another thing: there's a lot of competition in the car business. Even just here in Dallas there are probably six to eight guys who are as fast as Dennis and me. They're watching everything, too. So you've got to be quick. Don't be hesitant. If you see a car you want at a price point you can turn a profit on, trust your knowledge, trust your gut, and f—king yank it.

Your knowledge of cars is going to come from trial and error. There's no shortcut. There really isn't. So be prepared to invest some time and money, and don't be afraid to acquire that education over time. In order to be quick with your knowledge and your gut, you've got to put the time and experience in to develop that knowledge.

I played strictly with $8,000-to-$10,000 cars for the longest time just as a hobby on the side when I was younger and running the printing business. It was no big deal. I did four or five cars a year, maybe, tops. So when I decided to really start doing this, I gotta tip my hat to Dennis Collins. He would give me advice and steer me in the right direction. He's been in the business of buying and selling cars since high school. That's all he has ever done. He and I make it look easy now, but that's only because we've been doing it for so long.

Want some of my absolute best advice? Start out slow. Save your money. Do four or five car deals and save those profit dollars until you've doubled your money, then move up to the next level. For instance, let's say you decide you're gonna play with twenty-five grand. So play with that twenty-five grand until you've made twenty-five grand in profits. Then put your original twenty-five grand back in your bank and play with the money you made. Then? Keep compounding. If you're smart, if you focus on making that 15 percent or whatever number it is you stay focused on, if you don't get killed in a single deal but instead try to make money on average across all of the cars you buy and sell, it'll add up.

Think about it this way: if you add twenty-five grand to your bottom line every year, in four years you'll have made yourself $200,000. If you get yourself to a point where you have $200,000, you can go ahead and flip cars and make an additional $200K a year. If you get to $500,000 sitting there, ready to go, it is my opinion that you can flip cars constantly and make a million a year. Not everybody can do that well, of course, but that is definitely a possibility. I know because I've done it. I know because we're doing it here at Gas Monkey.

Here's the other big secret of my success. I'm talking long-term, lifelong success here, which is what I hope you really want. The trick is, do it right. Stand behind your s—t. Don't be shady.

There are a million shady car dealers out there, and there's a reason people think of used-car salesmen as the stereotypical shysters and jerk-offs who are just robbing you blind every chance they get. Don't be that guy.

Right from the start, with my very first flip as a teenager, I've told people exactly how it is. Don't try to hide a flaw or a defect. Don't try to hide the truth about the vehicle you're trying to sell. There are buyers out there for every car. They might not be buyers who are willing to pay whatever pie-in-the-sky number you think a particular car is worth. But there are buyers. There are buyers who want project cars, who want something that's dirt cheap just so they can mess around with it in the garage or fix it up as a long-term project. The marketplace will establish what the price is for any given car, and you've just got to accept that. Don't try to sell somebody something that's far less than what they think it is. Be honest. Be up-front. In the long run, that will bring you more customers—and repeat customers. And best of all, it just makes you and them feel good. Never forget: karma's a bitch!

In the eleven years I have been flipping cars as an actual business, I have bought back three cars from unsatisfied buyers. Three. That's it, out of hundreds and hundreds of cars. Well, guess what? I turned around and resold every one of those cars for more money than I sold it for in the first place, all without putting another dime into the car.

If you're flipping cars, every once in a while you'll get a guy that'll call you up after the sale is done and say, "Oh, you didn't say it was this and this and this," and all they're really trying to do is get you to return some of their money. They think they'll be able to guilt you into giving them a thousand dollars back or something.

I had one guy just recently who paid, I don't know, let's call it ten grand for some car I was selling. He called me up a few days later and was like, "Well, it's gonna need at least two thousand dollars' worth of work."

My response? "Well, sir, I feel like I described it correctly."

"Maybe so, but I'm not happy. I think you owe me that two grand. So why don't you send me two thousand dollars back and we'll be good."

"That's not how it works," I told him. "Look, I want you to be happy. If you're not happy, you're not happy. So I'm gonna send a tow truck over. They'll be there in an hour and I'll give your ten grand back today."

"Oh, man! No, no. I don't want to give up the car."

Another guy bought a car from me and took it to another state before he tried the same tactic. I knew I didn't want to tow the car anywhere. It would've cost me $1,800 in shipping. So I called a local storage lot and had it picked up, and I told the guy exactly what I was planning to do: "Dude, I'm gonna put it back on eBay, and you watch this."

So I sent him his money back, picked up the car, put it in a storage lot, put it back on eBay, and I sold it for $4,000 more than he paid for it. He calls me up and he goes, "Well, that's just f—king bulls—t."

I said, "What do you mean?"

He goes, "Well, you owe me four grand!" He wanted the whole four grand! He goes, "That's *my* profit."

I said, "Motherf—ker, you said you didn't like the car and you were unhappy. You have your money in your pocket and I picked my car back up. You're not entitled to s—t!"

You have to watch out for real crazies, too. For instance, we had a green Mustang that we put on eBay right after we featured the car on *Fast N' Loud*, and it went for a bid of $207,000. The deal ended up not happening because it was just some idiot being stupid on the Internet. He didn't have that kind of money and it was never going to happen. He wasted a lot of people's time and energy over nothing.

Long and short, be careful. Don't be stupid. Don't let people take advantage of you or try to rip you off. Especially online, you need to verify everything and be sure you're dealing with real people, real money, and real vehicles.

Dealing with all of that sort of s—t isn't for everyone. You've got to be tough, and fair, and determined, and willing to work hard and put up with a lot of BS to make money in the car business. There are a lot of people out there who lose their shirts selling cars! Don't be one of them. I don't want to hear people complaining, "Oh, Richard Rawlings said flipping cars was an easy way to make a living, and now I'm broke." I did *not* say this was easy! If you've read this book, hopefully you understand how much time and effort and blood and sweat went into building Gas Monkey Garage, and a whole lot of that time and effort was spent learning how to flip cars, going all the way back to when I was a teen. If you think this is gonna be easy, go back to the beginning and read it all again.

Is it fun? Hell, yes! It can be crazy fun! But that does not mean it's "easy."

Whether you're jumping into it as a little side hobby or thinking about making it a full-time job, flipping cars is a business. The only thing I can guarantee is this: if you're a true Gas Monkey type of individual and if you're willing to put in the work, you'll have a whole lot more fun flipping cars than doing just about anything else you can think of with that stash of cash that's sittin' in your bank account.

FLIP TIPS

To sum up, here are a few things to think on as you go about your flipping business. (My thanks to Tony and Aaron for chiming in on some of these, too!)

Don't limit yourself to your local area.

There are car bargains to be had all over the place—but they might not be where you're located. You'll notice on the show that we drive up to Michigan and Missouri and all sorts of places to find the diamonds in the rough we're looking for. The fact is, sometimes a car that's worth only $2,000 in Michigan will sell for $6,000 in Dallas, Los Angeles, or Miami. Know your markets, know what cars are going for in different parts of the country, and cast a wide net in your eBay searches.

Craigslist is a little more focused in that it's local, so it may take more work than eBay. If you want to find cars in the Dallas area, you go to that site. If you want to cast a wider net you have to look at Craigslist St. Louis, or Craigslist Atlanta, and so on.

One of Tony's tricks is to use a site called SearchTempest, and a subset of it called AutoTempest, that will search all of those sites and the broader Internet all at once. SearchTempest allows you to put in search terms for exactly what you're looking for. You can modify the search to weed out stuff that's more than a certain distance away, or to only find stuff that's close, or only stuff that's a certain price. Or, if you've already looked at eBay, you can eliminate eBay from your search. It's pretty rad.

Then, of course, you can pull away from technology entirely. There are a lot of people out there who don't use the Internet, who don't have a digital camera, who wouldn't know how to list a car on eBay if their lives depended on it. So if you really want to get dedicated you can search through the local classifieds for the local papers, but just like with Craigslist, you have to look in each city. Houston, Austin, Dallas, wherever it may be. Some papers list their classifieds online, but some of the smaller ones don't. And you never know what might pop up in some local classified somewhere. The best thing you can do is keep your eyes open. Don't be afraid to spread the word, too. If you're looking to buy cars, let your friends and family know so they'll keep their eyes and ears open, too. They might not know what kind of vehicle you're looking for or how to tell a gem from a rock, but that doesn't matter. As long as you're hearing about it when a car comes on the market, you can use your skills and knowledge and gut to know if it's something you want to look into or not.

You also have to consider shipping costs if a vehicle is located far away. If you're looking to buy a $3,000 car and it's $1,000 away, you really need to think about whether it's worth it or not.

You can also use that as a negotiating tactic. "Hey, I like your car. But it's a thousand dollars away and you're asking too much already."

Don't be judgmental or afraid of any particular location, though. Honestly, some of the cleanest cars we've found here at Gas Monkey have come from areas where you don't expect to get clean cars. You think of a place like Michigan as being terrible on cars, because of the harsh winters and the road salt that'll rust a car out twice as fast as it would rust out anywhere else. What you might not think of is that those cars might have been kept in a garage and not driven in the winter. Especially the fun two-door sports cars that we like to pick up and flip the most.

One of the cleanest little Comets we ever got was out of Detroit. It sat in a barn and only had like six thousand miles on it. You could look at it and put the story together: it looked like a car somebody bought for his wife to run to the store and she was simply not driving in bad weather. So it sat around and was driven very little, and never in bad weather, and all these years later it was a great car. It was absolutely original. No rust. Original hubcaps, tires . . . six thousand miles!

On the flip side, when it comes to selling a car, don't even limit yourself to staying in the country. The beauty of eBay is it goes out to the whole world, and there are certain types of cars that people in Europe are just clamoring to get their hands on.

With eBay, you throw up some pictures, a good description, and it goes around the world. Looking back on it as we're writing this book, out of the last ten Model As we've sold, six went to the UK, one went to Germany, and one to Finland. They can't get enough of them!

Get on the freight train.

You might wonder how in the heck people are affording to purchase cars from overseas in terms of shipping costs alone, but I'll tell you this: it can be cheaper to ship a car from Dallas to London than it is to ship it from Dallas to Los Angeles. It only costs maybe $300 to get it to the port here, and then the car goes into a container and onto a ship that won't get there for six to eight weeks—but it's cheap. To get a car to L.A., you have to pay for a driver and a proper transport. If you're going on an open transport and lumping it in with a bunch of other cars, it might be cheaper to get it to L.A. But in general, on a car that's worth $10,000, the Londoner's going to get the better bargain in terms of shipping costs.

When you're shipping a car in from a long distance, be sure to shop around for the best transportation price. There are online services such as uShip where you can enter in what you need shipped, and when, and where, and various individuals and companies will bid on the job. That can work out pretty well if it's just a one-time or occasional thing.

Here at Gas Monkey we use a professional auto-transport broker. Basically, there's a woman Tony can call up anytime and say, "Hey, we've got a Mustang at this address in Minneapolis. How much is it going to cost me to get it shipped down here to Dallas in the next week?" And she'll give him an estimated price. She then puts it out to her various transportation companies at that price and sees what she can find. Ideally, what you want to find is a truck that's already coming through that way, but they have an empty slot they can fill with your car. That's gonna be the most cost-effective way to move a vehicle from one part of the country to another, by far. And all of those costs affect your bottom line, so if you can save a hundred bucks here or there, it's worth it.

Now, occasionally when you're getting something from way up in Minneapolis or especially Detroit, it's a real chore to find someone to do the job. In those cases, if nobody bites at the price you'd like to pay, you have to raise your price by $100 or something in order to entice some trucker to go out of his way.

Any way you look at it, using a broker or a bidding service of some kind is going to be a huge savings over hiring some individual tow truck or flatbed driver to go and fetch a vehicle for you.

Don't be afraid to ask for more pictures.

In the Internet age, buying a car unseen is something you're going to have to do at some point. Here at Gas Monkey, we do it all the time. It's scary at first. It really is. But there are some things you can do to make it less scary.

First of all, it's possible to make almost anything look good in a picture or two. The right lighting can cover up all the blemishes and flaws. Super-savvy Internet users know how to use Photoshop to clean things up, too. So you've got to be careful. We have had a few cars, luckily not too many, but we have had a few cars that weren't as nice as they looked in the pictures. They come in and the paint's a little rougher than you thought it would be, or there's some hidden rust that needs major repair work.

Make sure you open up the pictures as big as you can, and really look 'em over. If you find something that looks fishy, contact the seller and ask them to send you some close-up shots of that fender or whatever it may be. Just be diligent. It's sort of like having a car inspected before you purchase it in person: if you find something's wrong with it, it doesn't mean you don't buy it. It just means you go, "Hey. This car has this and this and this wrong. You're asking fifteen grand, but I've got to fix all this. So it's only worth twelve."

Do your research when assessing value.

The most important thing is to make sure you're buying at the right price. Everything has a right price, and I mean that. There are junkers that are valuable as parts. Rusty pieces of old '50s fenders and doors can be cut up and marketed as decorations to sell for people's bars and man caves and shops, if that's what you're into. That's not a business I want to be in, but there's truly some kind of value in everything, and it's up to you to know what that value is.

For the older cars that we deal with at Gas Monkey, there's really no easy way to assess value, either. There's no Kelley Blue Book or Edmunds valuation that's going to keep up with the marketplace or even mention stuff like a Model A or a '32 Ford three-window.

Your best bet is to just keep track of what's selling in your area. Look on eBay, not for the listed price, but for the sold price. If some guy has a car on there listed for $50,000 and it doesn't have any bids, it's a pretty safe bet it's overpriced. If a similar car is getting lots of bids and it's hovering around twenty-five grand, then you're looking at a pretty accurate read on the marketplace most of the time. If you can find a bunch of closed auctions for a particular type of car, and there's an average price that they've all sold for, that's about as accurate an assessment as you're gonna get. Then all you have to do is compare and contrast: Does the vehicle you want to buy or sell have better brakes? Is it in better condition? Does it have special features, and are those features desirable? What's holding it back? Is there a must-do $2,000 repair or replacement? Because if so, you need to build that at least partially into the value you're assessing for sure.

One way to familiarize yourself with the marketplace, and with particular models of cars in general, is to go out and look for clubs and groups that are dedicated to those vehicles. All it takes is a Google search for "Model A enthusiasts" or "Gremlin fanatics" or whatever it is you're into and you're sure to come up with something. Just like everything else on the Interwebs, though, beware the phonies. If you're looking for particular information about values, or about parts, or about fixing something, don't just go to one message board and take somebody's word for it. A lot of people who post on message boards are full of s—t. So if you find an answer to a question you have, dig deeper. Check your sources. Go to two or three other message boards for answers, too. It's kind of like good reporters used to do in the olden days of newspapers: if you can corroborate your information from at least three different, unrelated sources, you can walk away knowing that the information you've got is probably true and accurate.

I've said this before, but just remember that there aren't any real shortcuts in this business. You've got to be diligent, and even then you're gonna make some mistakes. That's okay. Be prepared for it and it won't bother you. The way you pay for your education is through mistakes.

Learning to assess values is a process. And no matter how good you get at it, you can't make money on every car out there. That's rule number one. Not every car out there will make money.

Go ahead, repeat yourself.

If you find a certain car that works for you, a certain niche that makes you a tidy profit, write it down. Repeat it. See if you can get lightning to strike twice, and then three times, then four. Some sales are flukes—you simply find a buyer who's willing to pay too much for whatever car it is you're selling. That's fun when it happens, but the real profits come from finding a particular type of car you can get to know inside and out, and then making a profit off of every one of them that you flip. If you find something you can consistently and repeatedly make money on, stick with it and ride that ride for as long as you can, whether it's Mustangs or Model As or Ford pickups. What it is doesn't matter. If you find something and it works, ride it.

Make your losses work for you.

You can't win all the time. The difference between success and failure is learning how to make your losses work in your favor. I've already talked about not hanging on to a loser of a car and sinking tons of money into it thinking you're gonna turn a profit. You've got to drive that lesson into your head as much as possible if you want to make it in this game. Like I said, it all comes down to that famous Kenny Rogers lyric: You've got to know when to hold 'em, know when to fold 'em.

At one point before the show launched, I had about fifteen cars that I'd been holding on to and trying to sell. They were of all different makes and models. I had a few hundred thousand dollars tied up in them. I didn't have any cash freed up to go buy other cars. There was an auction coming and I finally said, "You know what? F—k it. Win or lose. I don't care. I've got to free up as much cash as I can, because cash is what allows me to make more money!"

Cash is my inventory, at the end of the day. So I took all those cars to one auction and proceeded to have quite a few beers as one after another lost money.

I lost about $120,000 that day. That's more than most people in this country make in a year.

Remember, though, I've always kept a low overhead, and I've always been smart with my money (just like my shop teacher taught me). So this wasn't a devastating "loss." Was it disappointing? Yes! But in the long run, it wasn't a loss at all.

By selling all of those cars at once, I freed up about two hundred grand. Having that cash allowed me pay the rent and other expenses, and then to go buy and sell a bunch of cars and start making some money again.

The trick is, I was able to put a lot of cash together all at once as opposed to just selling one car and losing some money (and then all that money is gone), and then selling another car and losing that money (and then all of that money is gone, too). I just covered all of them, took my loss, and then I had a giant lump sum of cash with which I could go buy ten or fifteen cars that would make money. So what to some people looked like a great big loss was actually a great big win!

Obviously you're not going to be flipping hundreds of thousands of dollars' worth of cars anytime soon if you're just getting started. But the principle is the same. If your cash is all tied up in vehicles that aren't moving, sell those vehicles and reinvest whatever cash you can into something that will actually make some money.

Stay on the up-and-up.

I've already mentioned treating your customers right. Don't cheat 'em. Don't trick 'em. Honesty is the best policy. Karma is a bitch. All of that is important.

Here's another bit of advice: don't f—k with the IRS.

It's easy for people to get into flipping cars for cash without keeping track of it all and reporting those sales to the government. Fair warning: you do that at your own peril. If you have any kind of success flipping cars, the IRS will find you. The fact is there are title transactions and registrations and all sorts of things that make cars and car sales trackable by all sorts of state and government entities. Don't f—k with any of them.

If your town or county or state requires you to get a permit to buy and sell cars, go get the damned permit! It's much less of a headache and much lower of a cost than dealing with fines and all sorts of problems after the fact. Believe me. Pay your taxes. Pay your fees. Read up on the local laws that pertain to vehicle sales. If you're going to do a lot of this, take the time to set up an actual business, like an LLC (a limited liability company) or something. There are companies all over the Interwebs that'll help you do that. Or spend a few bucks and hire a local business attorney to walk you through it. Yes, there's some paperwork. Yes, it's a pain in the ass. But again, it's far less paperwork and far less of a pain in the ass to do it up front than it is to get in trouble, get sued, pay a fine, or—God forbid—go to jail for doing things the wrong way.

Education is expensive!

Here's a little something to chew on, courtesy of Aaron Kaufman:

> *Education is expensive. You go, "Sure. Yeah, I remember my college was expensive." I didn't say "college." I said education is expensive. Whether you're going to soak it up off somebody else or you're going to take the hard-nosed route and learn it yourself, education comes at a price—and the education brought to you through hard luck tends to be some of the best lessons learned.*
>
> *Certainly trying to buy and sell cars is not going to be an easy thing for somebody to get some education. It's a very expensive education, and I'd warn anybody who's thinking about it to make sure they add up all the costs of doing business before they get too deep into it.*
>
> *On every single car you're considering, you have to ask yourself: by the time you put a battery in it, fuel in it, move it, transport it, make the phone calls, and not do your other job that you're going to potentially get fired from for trying to moonlight with this flipping-cars business, are you really going to make any money?*

Aaron! Dude! Why you gotta be so negative? Yes, it's important to consider all of those factors. You have to look at the cost of everything in life to know if it's worth it. And, yes, you need to realize that it's going to take time and money to get educated in the field of flipping cars. I think I've made that point pretty clearly. Just be smart about it. Don't risk more money than you're able to comfortably risk. Start slow. Take it one step at a time. Educate yourself. And then really think about why you're doing this. If it ain't fun, if you don't love it, if it's not making money, don't do it. That's pretty much the Gas Monkey golden rule right there!

Beware of runaway trends.

One way to make some really quick money in the car business is to jump on a trend. All of a sudden a certain make or model will start going up in price like crazy. For some reason everybody wants a certain old car that nobody else wanted for years and years. You buy one cheap, and a month later you sell it for 300 percent more than you paid for it. It's awesome!

The problem comes when you're trying to chase those trends all the time. It's easy to get caught up, and go broke.

A few years ago, every guy out there wanted to get his hand on an old muscle car with a Hemi engine. Everything was Hemis! Cars that were worth $50K were suddenly selling for $100K, then $200K, then $400K at major car auctions. It was insane! Then, all of a sudden, the market was saturated. Everybody who wanted and could actually afford a Hemi had one. *Crash!* All of a sudden, you couldn't sell a Hemi for $100K no matter how hard you tried, which means a whole bunch of people out there got stuck with cars they're gonna lose $300,000 on.

Don't be one of those guys.

The fact is, sometimes the marketplace gets rigged. There are groups of guys who buy up certain types of cars, and who'll bid them up at auctions just to make it look like the cars are trending upward as a whole. Then all of a sudden people start buying them, thinking that they're noticing a real trend. They'll get caught up spending too much and eventually lose their shirts trying to unload those cars when they realize there really wasn't any type of a broad-based buying frenzy going on at all.

Auctions can be tricky. It's a big world full of big money, and unless you know what you're doing, you can get burned. You can occasionally find a real bargain at an auction, so I'm not saying to skip them. Just be very well educated and know your limits before you go into one.

That said, it is important to try to watch for certain trends and cycles that come around. Cars go in and out of favor all the time. For years, nobody wanted Ferrari 308s—you know, the *Magnum P.I.*–style Ferraris. They'd gone out of favor. You could pick one up for like twenty-five grand! A Ferrari! But then you wouldn't be able to unload it. Now? All of a sudden they're coming back into favor. People like the look of 'em again. At one point it looked like we were entering into a nice upward trend. So I started buying them. It cost me more than fifty grand to pick one up that was in decent condition and didn't need any work. But I know these markets, and I'm pretty sure I'll be able to off-load that 308 at a $20K or $30K profit in just a matter of months. The reason I can say "I'm pretty sure" is because dealing with cars is all I do now. I live it and breathe it. I know the guys who buy these things and who are looking for these cars. I see the trends because I'm right in the thick of it. Fifty grand is a lot of money to risk, and I'm fortunate enough to have that sort of money to risk nowadays. That is not the case with most people.

So instead, stick to cars you know, in a price range you can feel comfortable with. At Gas Monkey we focus on fun cars, two-door sports cars and hot rods that people buy because they think they're cool. We never sell four-doors at Gas Monkey. No matter how great of a bargain some four-door might look like on paper, we always lose money on 'em. So we just don't touch 'em anymore. Of course, people associate Gas Monkey with cool cars, so maybe we've created our own monster in that realm. Maybe four-door cars would sell just fine wherever you are. Maybe you could even develop a niche selling four-door family sedans! I don't know. Anything's possible. You're much better off sticking to the cars you're passionate about and that consistently make money if you're gonna make this work in the long run.

For instance, trucks are pretty much what people look for in the hot-rod world right now, but we'll only buy short-wheelbase trucks here at Gas Monkey. We won't buy a long wheelbase, and primarily we buy what they call fleet sides. There are a lot of cool step sides out there, but there was a trend probably fifteen, twenty years ago where everybody was building step-side trucks. So there's plenty of them out there. But nobody was doing fleet sides because that wasn't cool then, so that's what we look for now.

When it comes to tastes in vehicles, it's sort of like tastes in fashion: almost everything is cyclical. If you wait long enough, certain trends will come back around. Buying a car isn't like buying a leather jacket, though. You don't want to get caught having to wait twenty years before a vehicle you purchased comes back into style!

To give you some more insight, here's Tony's two cents on the matter:

The thing about a crystal ball is if you had one, it would be priceless. It's hard to guess what's gonna happen in the next three to six months based on what has happened in the last three to six months. That's not always accurate. There are a lot of times that can be deceptive. If you do want to jump on trends, you have to know what you're doing, and you've got to be quick. If you start to see old Ferraris go up, old Porsches, old Mercedes, you move on those and you have to move quickly to get them. But then you have to get rid of them at the right time, too.

Keep an eye on the broader economy as well. Back in 2007 and 2008, there was an actual fault line in the market with the economy doing what it did, because these cars, at the end of the day, they are wants. They are not needs. These are things people play with and enjoy, hopefully. A lot of people don't, but hopefully you can enjoy an old car and owning it and working on it and appreciating it, and then on the day it sells, hopefully you enjoy that day, too.

If you're flipping on a larger scale, it becomes a numbers game. If you buy ten cars, hopefully you make money on eight of them, and you expect to lose money on two of them. Now, if that number switches to only making money on six and losing on four, then you are not doing real good. It would be great to get to a point where you are not losing money on any of the cars you buy, but as long as you are buying enough there's a number where it makes sense. Say if you only buy five cars and make money on all five, you might make a certain percentage of profit. But if you're buying ten and only losing money on two of them—and making money on the other eight—you're actually making more money than you would've if you'd just stuck to the five moneymaking cars. That's how the big dealerships think, too, except they're selling hundreds of cars, knowing they might lose money on twenty or thirty of them. They're still making far more money that way than if they were selling fewer cars.

There's also a saying that holds true in the car-flipping business: You make your money when you buy the car, not when you sell it. If you buy a car for seven and you know it's worth nine or ten, then you know you've made money the moment you bought it. If you know it's worth ten and you're paying eight or nine, you might be in trouble. Be smart, do your research, know your market, and don't overpay at the outset. Then you're good! And remember, it doesn't always have to make sense to make money. That's something I've definitely learned here at Gas Monkey. Just because you don't like a certain car doesn't mean somebody else won't. Just because a certain type of car isn't a good car doesn't mean there aren't buyers who love to buy 'em because of the way they look or whatever the trend may be. All you need to know is that people are buying them, and that in the end, you made money.

Find a mentor.

I've said it before, but I owe Dennis Collins a huge amount of gratitude for teaching me the ins and outs of flipping cars back when I first launched Gas Monkey Garage. He'd been doing it full-time since high school, and he'd already learned so many of the lessons I'm sharing here with you now—only he had to learn 'em the hard way, just like I did. So not only was he a great friend, but he was a mentor to me. He truly helped me refine my skills in a very short period of time: he was right there helping me make decisions on a lot of purchases, and plenty of the sales, too.

The thing about having a mentor like Dennis is I knew I could trust him. There were a lot of other guys I knew in the car business who I was friendly with, but I'd call 'em up to get their help, like, "Hey, man, what do you think a '63 Corvette split-window in such-and-such a condition is worth?" and they would be like, "Well, I don't know, man." They'd come around and throw out a price and say, "Hey, if you get it for that, bring it to me and I'll give you a thousand dollars' profit just to take it off your hands." Come to find out that car was worth fifteen grand more than they offered to pay me! It was all part of the game.

With Dennis, there was no game between us. He'd steer me right. Always.

Beware of friendly car guys who'll take advantage of a newbie. I ran into a lot of them. And while it's tempting to go and make a quick thousand or two thousand bucks flipping something to an eager buyer who's going to flip it again, you have to be smarter about it and realize that making $1,000 is pretty stupid if what you're really doing is giving up $10K.

Having a mentor and friend in Dennis was huge for me then, and it's huge for me now. He's the kind of friend that if I were in a tough spot, I could call him up and say, "Dude, I need one hundred grand right now," and he wouldn't even ask me what it was for. All he would say is, "Come get it." All in all, Dennis made all of his money the same way that I made my money: by hustling. I mean, he's in the Jeep business primarily, and he's the largest reseller of used Jeeps in America. It's been a very lucrative business for him. But that doesn't mean he rolls over and goes to sleep and sips margaritas all day on the beach. He's still hustling like me because it's fun! Flipping cars, building hot rods, racing in rallies . . . for Gas Monkeys like us, that s—t never, ever gets old.

If you're lucky enough to find a mentor like Dennis in this business, you almost can't help but be successful. Listen to what they tell you. Take their advice. Put it to good use. You'll be all the better for it.

Oh, and P.S.: having a good supportive wife helps, too. The fact that Sue didn't try to stop me from launching Gas Monkey in the first place, even though she doubted what I was doing, was a real bonus for me in the early days. The fact that she put up with my weeks on the road, and the endless long hours and money problems and everything else, was incredible. The fact that she remained my friend even after we'd divorced, and was there for me at the very moment when the TV show I'd dreamed about finally came my way, meant the world to me. In fact, if it wasn't for her support, you might not know who I am, and I might not be writing this book!

If you've got a good woman backing you up, don't lose her, man.

Just as this book was getting ready to go to print, I flew down to Cabo to surprise Dennis on his birthday. Sue and I had been hanging out a lot in the weeks and months before that trip, so she came along, too. There I was, toasting my best friend and mentor, when I decided it was time to do right by my other best friend in the world, too. Right then and there, on that weekend trip south of the border, Sue and I got hitched again.

I'd tell you more . . . but that story could fill up a book all on its own.

AND IN THE END . . .

Well, I guess that's it. For now, anyway.

I suppose I should come up with a good ending here, like one of my favorite flicks from the 1980s might have. Hmm . . . I got it! How about a moral-of-the-story type of ending?

The moral of this story is that after all's said and done, after all of the rebellious cross-country adventuring and bucking the system and making my own way in my life and career, what I really did is pretty simple: I followed my dad's advice. I found a good-paying job with great benefits and a great retirement package. It's just not what most people think of when they think about that kind of a "job." Instead of clocking in at some nine-to-five behind a desk and working for somebody else's profit, I'm clocking in for however long it takes to work to create my own pay. I've got all the benefits that come from doing something I love. And I'll be buying myself my own gold watch when it's time to retire, if that time ever comes.

I'm not relying on anyone else to fulfill my destiny. I lean on others and I benefit from the expertise and skills of the amazing Gas Monkey family around me, for sure. I'm fortunate to have all of the Gas Monkeys in my life. But in the end, my success is up to me. And so far, I'm kicking ass!

The Gas Monkey empire is growing like crazy, and trust me, you'll flip out when you see what's coming in the next few years. My plans are gigantic, man! Bigger than anyone (other than me) can imagine.

You know what's really exciting, though? In the end—and I'm talking the real end here—none of this is really about money. I love money, don't get me wrong. But what I want to build through all of my Gas Monkey endeavors is the legacy of a massive, unstoppable machine. A machine that gives back. A machine that can provide for those who can't provide for themselves. The legacy of all of this crazy success I'm building is that I want to be able to take the opportunity that I've been given and to give it all back.

That's exactly why I launched the Gas Monkey Foundation. You can check it out online, but basically what we're doing is taking donated cars from people all over the country and selling them at auction to make money for this nonprofit organization I set up. Donating cars is easy. People fill out a form online, and we send someone to pick up their old used car (no matter what condition it's in), and we give them a donation receipt for the value of the car so they can save some serious money on their taxes. The Foundation then donates the money to established charitable organizations that will put the money directly to good use. I'm aiming to raise millions of dollars with this endeavor to put into Alzheimer's research, protecting the environment, saving wildlife and fighting human trafficking. I want to use this fame and notoriety that Gas Monkey and *Fast N' Loud* have built to do all sorts of good things in the world. And I'm doing it through the buying and selling of old cars! What could be better?

As I sit here looking around at all of these people who are benefiting from and enjoying the brand I've built, I know I've achieved something great in life already. So honestly, I don't care if I die broke. As a matter of fact, if you've got to go you might as well spend every penny you've got and have one hell of a party of a life, am I right? To me, the greatest feeling of all will be knowing that the money I've made and the businesses I've built have gone on to help a hell of a lot of people.

Just as long as I can afford to die with a bottle of Miller Lite in my hand, I'm f—king good.

So, until next time . . . Rock on, people! *Whooo!*

ACKNOWLEDGMENTS

This book would not have been possible without all of my Gas Monkey fans. So first and foremost, thank *you*!

Thanks also to Aaron Kaufman, who joined in on this unpredictable ride when all I had was a crazy idea and a dream. He's been living that dream with me every day since. This book also couldn't have come together without the assistance of a whole bunch of really great people, including the crew of Gas Monkeys who kept things going at the garage while I took the time to get it done. Specifically, there is no way this book would have happened without the above-and-beyond efforts of Christie Brimberry and my sister, Daphne Kaminski; not to mention Lauren Parajon, who helped a lot with this book.

I want to thank my editor, May Chen, and the whole team at HarperCollins, who showed their support for this book from the start. My literary agent, Cait Hoyt, for taking care of business and hooking me up with such a great team. My coauthor, Mark Dagostino, for traveling to Dallas and helping to turn my rambling memories into the words you see on these pages. And to everyone at Discovery for helping to move my *Fast N' Loud* story from the screen right into your hands. I also want to thank my manager, Antranig Balian, and my television agent, Hans Schiff, for always having my back.

Big thanks to Craig Piligian, president and CEO of Pilgrim Studios, for his perseverance and dedication to *Fast N' Loud*, and for getting this thing sold to Discovery. I also want to thank the people at Pilgrim who work on the show daily, most notably Eddie Rohwedder and Rebecca Graham-Forde.

I would like to thank the inventors of Miller Lite for making the end of my long workdays a whole lot better.

And I especially want to thank my dad for helping to make me everything I am today.

All of these things were given to me by God, so I'm thankful for the crazy life and family God has blessed me with.